# A MANUAL OF SHOEMAKING AND LEATHER AND RUBBER PRODUCTS

WILLIAM H. DOOLEY

Published by Left of Brain Books

Copyright © 2021 Left of Brain Books

ISBN 978-1-396-31937-2

*First Edition*

All rights reserved. No part of this publication may be reproduced, distributed, or transmitted in any form or by any means, including photocopying, recording, or other electronic or mechanical methods, without the prior written permission of the publisher, except in the case of brief quotations embodied in critical reviews and certain other noncommercial uses permitted by copyright law. Left of Brain Books is a division of Left Of Brain Onboarding Pty Ltd.

# Table of Contents

| | |
|---|---:|
| PREFACE | 1 |
| CHAPTER ONE. Fundamental Shoe Terms | 3 |
| CHAPTER TWO. Hides and their Treatment | 5 |
| CHAPTER THREE. Processes of Tanning | 13 |
|     Leather for Belting | 31 |
|     Rawhide Products | 33 |
|     The By-products of a Leather Belting Factory | 34 |
|     Round Belt Making | 34 |
|     Properties of Tanned Leather | 35 |
|     Substitutes for Leather | 36 |
| CHAPTER FOUR. The Anatomy of the Foot | 38 |
| CHAPTER FIVE. How Shoe Styles are Made | 46 |
| CHAPTER SIX. Departments of a Shoe Factory— Good-Year Welt Shoes | 51 |
| CHAPTER SEVEN. McKay and Turned Shoes | 76 |
|     History of the Turn Shoe | 83 |
|     Standard Screw Shoemaking | 84 |
| CHAPTER EIGHT. Old-fashioned Shoemaking and Repairing | 86 |
|     How Shoes are Repaired | 88 |
|     Shoe Repairing | 88 |
|     Modern Method of Repairing Shoes | 90 |
| CHAPTER NINE. Leather and Shoemaking Terms | 93 |
| CHAPTER TEN. Leather Products Manufacture | 111 |
|     Automobile and Furniture Leather | 113 |
| CHAPTER ELEVEN. Rubber Shoe Manufacture | 115 |
|     Rubber Heels | 123 |

Chemistry in the Manufacture of Rubber Goods                126
Rubber Terms                                                126
**CHAPTER TWELVE.** HISTORY OF FOOTWEAR                     129

# PREFACE

THE author was asked in 1908 by the Lynn Commission on Industrial Education to make an investigation of European shoe schools and to assist the Commission in preparing a course of study for the proposed shoe school in the city of Lynn. A close investigation showed that there were several textbooks on shoemaking published in Europe, but that no general textbook on shoemaking had been issued in this country adapted to meet the needs of industrial, trade, and commercial schools or those who have just entered the rubber, shoe, and leather trades. This book is written to meet this need. Others may find it of interest.

The author is under obligations to the following persons and firms for information and assistance in preparing the book, and for permission to reproduce photographs and information from their publications: Mr. J. H. Finn, Mr. Frank L. West, Head of Shoemaking Department, Tuskegee, Ala., Mr. Louis Fleming, Mr. F. Garrison, President of *Shoe and Leather Gazette*, Mr. Arthur L. Evans, *The Shoeman*, Mr. Charles F. Cahill, United Shoe Machinery Company, Hood Rubber Company, Bliss Shoe Company, American Hide and Leather Company, Regal Shoe Company, the publishers of *Hide and Leather*, *American Shoemaking*, *Shoe Repairing*, *Boot and Shoe Recorder*, *The Weekly Bulletin*, and the New York Leather Belting Company.

In addition, the author desires to acknowledge his indebtedness to the great body of foreign literature on the different subjects from which information has been obtained.

An Old-Fashioned Shoemaker. *Frontispiece.*

# CHAPTER ONE.

## Fundamental Shoe Terms

BEFORE explaining the manufacture of shoes, it is necessary to fix definitely in our minds the names of their different parts. Examine your shoes and note the parts that are here described.

The bottom of the shoe is called the sole. The part above the sole is called the upper. The top of the shoe is that part measured by the lacing which covers the ankle and the instep. The vamp is that section which covers the sides of the foot and the toes. The shank is that part of the sole of the shoe between the heel and the ball. This name is often applied to a piece of metal or other substance in that part of the sole, intended to give support to the arch of the foot. The throat of the vamp is that part which curves around the lower edge of the top, where the lacing starts.

Backstay is a term used to denote a strip of leather covering and strengthening the back seam of the shoe. Quarter is a term used mostly in low shoes to denote the rear part of the upper when a full vamp is not used. Button fly is the portion of the upper containing the buttonholes of a button shoe. Tip is the toe piece of a shoe, stitched to the vamp and outside of it. The lace stay is a term used to denote a strip of leather reënforcing the eyelet holes. Tongue denotes a narrow strip of leather used on all lace shoes to protect the instep from the lacing and weather.

Foxing is the name applied to leather of the upper that extends from the sole to the laces in front, and to about the height of the counter in the back, being the length of the upper. It may be in one or more pieces, and is often cut down to the shank in circular form. If in two pieces, that part covering the counter is called a heel fox. Overlay is a term applied to leather attached to the upper part of the vamp of a slipper. The breast of the heel is the inner part of the heel, that is, the section nearest the shank.

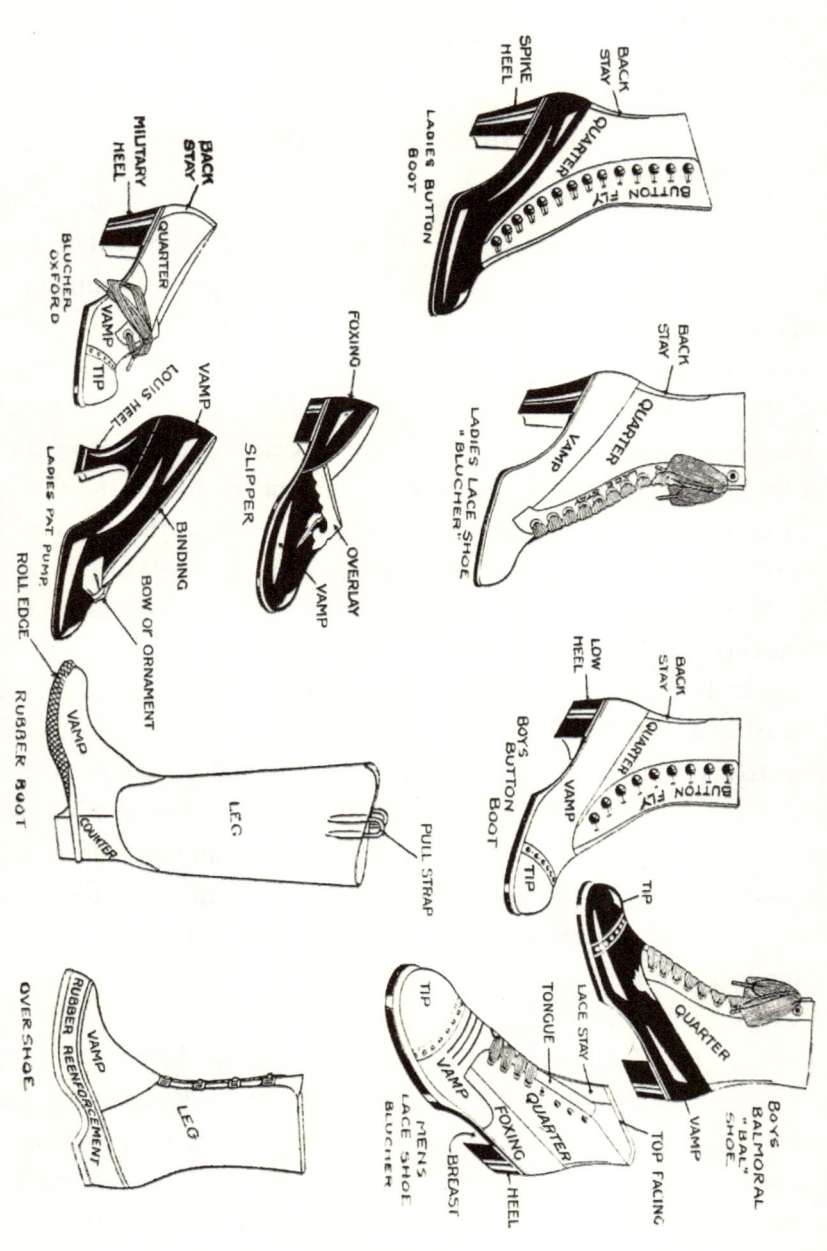

Names of the Different Parts of Foot Wear.

# CHAPTER TWO.

## HIDES AND THEIR TREATMENT

IF we examine our shoes, we will find that the different parts are composed of material called leather. The bottom of the shoe is of hard leather, while the part above the sole is of a softer, more pliable leather. This leather is nothing more than the hides of different animals treated in such a way as to remove the fat and the hair.

After the hides have been taken from the dead body of the animal, they are quite heavily salted to preserve them from spoiling. In this salted condition they are shipped to the tanneries.

The process or series of processes by which the hides and skins of animals are converted into leather is called tanning. The process may be divided into three groups of subprocesses as follows:—

Beamhouse process, which removes the hair from the hides and prepares them for the actual process of the tanning or conversion into leather; tanning, which converts the raw hide into leather; and finishing, which involves a number of operations, the objects of which are to give the leather the color that may be desired and also to make it of uniform thickness, and impart to it the softness and the finish that is required for a particular purpose.

Hides are divided roughly in the tannery, according to the size, into three general classes:—

(1) Hides, skins from fully grown animals, as cows, oxen, horses, buffaloes, walrus, etc. These are thick, heavy leather, used for shoe soles, large machinery belting, trunks, etc., where stiffness, strength, and wearing qualities are desired. The untanned hides weigh from twenty-five to sixty pounds.

(2) Kips, skins of the undersized animals of the above group, weighing between fifteen and twenty-five pounds.

(3) Skins from small animals, such as calves, sheep, goats, dogs, etc. This last group gives a light, but strong and pliable leather, which may be used for a great many purposes, such as men's shoes and the heavier grades of women's shoes.

The hides, kips, and skins are divided into various grades, according to their weight, size, condition, and quality.

The quality of the hides not only depends upon the kind of animal, but also upon its fodder and mode of living. The hides of wild cattle yield a more compact and stronger leather than those of our domesticated beasts. Among these latter the stall-fed have better hides than the meadow-fed, or grazing cattle. The thickness of the hide varies considerably on different animals and on the parts of the body, the thickest part of the bull being near the head and the middle of the back, while at the belly the hide is thinnest. These differences are less conspicuous in sheep, goats, and calves. As regards sheep, it would appear that their skin is generally thinnest where their wool is longest.

In the raw, untanned state, and with the hair still on, the hides are termed "green" or "fresh." Fresh, or green hides are supplied to the tanners by the packers or the butchers, or are imported, either dry or salted.

Hides are obtained either from the regular packing houses or from farmers who kill their own stock, and do not skin the animal as scientifically as the regular packing houses, in which case they are called country hides. There are different grades of hides and leather, and these different grades are divided in the commercial world into the five following grades:—

### I. Native hides

Native Steers
Native Cows, heavy
Native Cows, light
Branded Cows
Butts
Colorado Steers
Texas Steers, heavy
Texas Steers, light
Texas Steers, ex-light
Native Bulls
Branded Bulls

### II. Country hides

Ohio Buffs
Ohio Ex.
Southerns

### III. Dry hides

(Raised on plain. Rough side suitable for soles.)

Buenos Ayres

### IV. Calfskins (Green salted)

Chicago City

### V. Paris city calfskins

Light
Medium
Heavy

Hides obtained from steers raised on Western farms are known as native steer hides.

Native cowhide (heavy) is hide weighing from fifty-five to sixty-five pounds, obtained from cows.

Native cowhide (light) is cowhide weighing under fifty-five pounds.

Branded cowhide is hide obtained from cows that are branded on the face of the hide.

Butts is a term applied to the part of the hide remaining after cutting off the head, shoulders, and strip of the belly.

Colorado steer hide is from Colorado steers, which are very light.

Texas steer hide comes in three grades, heavy, light, and extra light. The heavy grade is very heavy because the animal is allowed to graze on the plains. That is the reason why it is heavier than the Colorado steer hide, which is raised on the farm.

Bull hide is divided into two classes, the regular hide and the branded grade. The branded grade usually is one cent a pound less than the regular.

Country hides are of three grades, Ohio Buffs, Ohio Ex., and Southern. The Ohio Buffs weigh from forty to sixty pounds. The Ohio Ex. weighs from twenty to forty pounds. Southern hides have spots without hair and other blemishes on them, due to the sting of insects. This makes the Southern hide inferior to the Ohio, Indiana, Michigan, and Chicago hides that have no such blemishes. Ohio Butt hides are the best, because in Ohio they kill a great many young calves, while in Chicago young cows (that have calved) are killed, causing the hide to be flanky.

The season of the year in which cattle are slaughtered has considerable influence upon both the weight and condition of the hide. During the winter months, by reason of the hair being longer and thicker, the hide is heavier, ranging from seventy-five to eighty pounds, and gradually decreasing in weight as the season becomes warmer and the coat is shed, until in June and July it weighs from seventy down to fifty-five pounds, the hair then being thin and short. The best hides of the year are October hides, and short-haired hides are better for leather purposes than long-haired ones.

A thick hide which is to be used for upper leather is cut into sides before the tanning process is completed. This is performed by passing it between rollers where it comes in contact with a sharp knife-edge, which splits it into two or more sheets. Great care must be exercised in cutting the leather in order to have good "splits" (sheets of leather). A split from a heavy hide is not as good as a whole of a lighter leather.

Butts and backs are selected from the stoutest and heaviest oxhides. The butt is formed by cutting off the head, the shoulder, and the strip of the belly. The butt or back of oxhide forms the stoutest and heaviest leather, such as is used for soles of boots, harness, etc.

Hides and skins, are received at the tannery in one of three conditions, viz. green-salted, dry, or dry-salted. Very few hides are received by tanners in fresh or unsalted condition, salt being necessary to preserve them from decay. Green-salted hides are those that have been salted in fresh condition, tied up in bundles, and shipped to the tanner. Dry hides are those that were taken from the carcass and dried without being salted; these are usually stiff and hard. Dry-salted hides are hides that were heavily salted while they were fresh, and then dried. The hides and skins that are received from the slaughterhouses of this country are almost invariably green-salted; those from foreign countries are green-salted, dry, and dry-salted.

Green-Salted Calfskin.

It does not matter in what condition the hides are received or the kind of leather into which they are to be tanned; they all require soaking in water before any attempt is made to remove the hair or to tan them. The object of the soaking process, as it is called, is to thoroughly soften the hides and to remove from them all salt, dirt, blood, etc. Ordinary hides are usually soaked from twenty-four to forty-eight hours. Dry hides require much longer. The water should be changed once or twice during the process, since dirty water may injure the hides. Soft water is better than hard for this process. Where the water is hard, it is customary for the tanner to add a quantity of borax to it to increase its cleansing power and to hasten the softening of the hides.

When dry hides have become soft enough to bend without cracking, they are put into a machine and beaten and rolled, then soaked again until they are soft and pliable. It is very important that all the salt and dirt are removed during the process of soaking, as they injure the quality of the leather if they are not removed before the hides are unhaired. When the soaking process is completed, the lumps of fat and flesh that may have been left on by the butcher are removed by hand or by a machine, and the hides are then in condition to be passed along into the next process. The parts that cannot be made into leather, such as tails, teats, etc., are trimmed off before the hides are soaked. Large hides are cut into two pieces or halves, called "sides," after they have been soaked.

For the purpose of taking the hair from the hides and skins, lime, sulphide of sodium, and red arsenic are used. Lime is sometimes used alone, but usually one of the other two chemicals is mixed with it. The lime is dissolved in hot water, a quantity of either sulphide of sodium or red arsenic is added to it, and the solution is then mixed with water in a vat, the hides being immersed in this liquor until the hair can be easily removed. The action of the unhairing liquor is to swell the hides, then to dissolve the perishable animal portion and loosen the hair so that it can be rubbed or pulled off.

There are several different processes of unhairing the hides. Each tanner uses the process that will help to give the leather the qualities that it should have, such as softness and pliability for shoe and glove leather, or firmness and solidity for sole and belting leather. This is one of the most important in the series of tannery processes, and if the hides are not unhaired properly and not prepared for tanning as they should be, the leather will not be right when it is tanned and finished.

There is also a process of unhairing, called "sweating," which softens the hide and loosens the hair so that it can be scraped off. In this process the hides begin to decay before the hair is loose; it is therefore a dangerous process to use and must be carefully watched or the hides will be entirely spoiled. Sweating is never used for the finer, softer kinds of leather. It is applied chiefly to dry hides for sole, lace, and belt leather. It is an old-fashioned process and is not used as much nowadays as some years ago.

The pelts of sheep are salted at the slaughterhouses and then shipped to the tannery. Here they are thrown into water and left to soak twenty-four hours to loosen the dirt and dissolve the salt. The pelts are next passed through machines that clean the wool, and any particles of flesh remaining on the inner or flesh side are removed. The pelts are then in condition to have the wool removed. As long as a sheepskin has the wool upon it, it is called a pelt; as soon as the wool has been taken off, it is called a skin or a "slat."

Each pelt is spread out smoothly on a table with the wool down and the inner or flesh side up. A mixture of lime and sulphide of sodium is next applied uniformly over the skin with a brush. The pelt is then folded up and placed in a pile with others. The solution that was applied penetrates the skin and loosens the wool, which, at the end of twenty-four hours, more or less, can be easily pulled off with the hands or rubbed off with a dull instrument or stick. The workman must be careful not to get any of the solution on to the wool, as it dissolves it and makes it worthless. Since the wool is valuable, the solution must be applied to the flesh side very carefully so that it does no injury. The wool that is removed from the skins is called "pulled wool."

The slat is now ready to be limed, washed, pickled, and tanned. Heavy skins are often split into two sheets after they have been limed. The part from the wool side is called a skiver, and that from the flesh side is called a flesher.

After the skins have been limed, they are bated and washed, which makes them soft, clean, and white; they are then put into a solution of salt, sulphuric acid, and water, called "pickle," and after a few hours they are taken out, drained, and tanned.

Large quantities of sheepskins are sold to tanners in the pickled condition by those who make a business of preparing such skins and selling the wool. Pickled skins can be kept an indefinite length of time without spoiling; they can also be dried and worked out into a cheap white leather without any further tanning whatever. Most of such skins, however, are sold to tanners, who tan them into leather. Sheep-skins contain considerable grease, which must be removed before the leather can be sold.

For some processes of tanning, calfskins, goatskins, and cattle hides are also pickled the same as sheepskins; for other processes they are not pickled, but are thoroughly bated or delimed, washed, and cleansed. Heavy hides are sometimes split out of the lime; more frequently, however, they are not split until after they have been tanned.

To capitulate, the preparatory processes may be briefly described as follows:—

Soaking, which dissolves the salt, removes the dirt and makes the hides soft and comparatively clean.

Liming and unhairing, which swell the hides and dissolve the perishable animal portion, loosen the hair, and put the hides into proper condition for tanning. Hides tanned without liming, even if the hair is removed by some chemical, do not make pliable leather, but are stiff and hard.

Bating, which removes the lime from the hides.

Pickling, which helps in the tanning later, and keeps the hides and skins from spoiling if they are not tanned at once.

The lumps of fat and flesh that may be on the hides are removed by machinery or by placing the hide over a beam and scraping it with a knife. The hair, when it is loosened by the lime, is removed by a machine or by hand.

# CHAPTER THREE.

## Processes of Tanning

THE various processes of tanning may be roughly divided into two classes, vegetable chemical and mineral chemical. The first class is often spoken of in tanneries simply as the "vegetable" while the second is called "chemical" process. In the vegetable processes the tanning is accomplished by tannin, which is found in various barks and woods of trees and leaves of plants. In the so-called chemical processes the tanning is done with mineral salts and acids which produce an entirely different kind of leather from that procured by vegetable tanning.

There is also a method of tanning, or, more properly speaking, tanning, in which alum and salt are used. This process makes white leather that is used for many purposes; it is also colored and used in the manufacture of fine gloves. Leather is also made by tanning skins with oil. Chamois skins are made in this way.

The materials that are used to tan hides and skins act upon the hide fibers in such a way that the hides are rendered proof against decay and become pliable and strong. There are many vegetable tans; they are used for sole leather, upper leather, and colored leather for numerous purposes. The bark of hemlock trees is one of the principal tans. The woods and barks of oak, chestnut, and quebracho trees are often used. Palmetto roots yield a good tan. Large quantities of leather are treated with gambier and various other tanning materials that come from foreign countries. Sumac leaves, which are imported from Sicily, contain tannin that makes soft leather suitable for hat sweat-bands, suspender trimmings, etc. Sumac is also obtained from the State of Virginia, but the foreign leaves contain more tannin and make better leather than the American.

To a large extent the so-called chemical processes have supplanted the vegetable processes, that is, old tan bark and sumac processes; but in some tanneries both methods are used on different kinds of skins.

In the old bark process the tan bark is ground coarse and is then treated in leaches with hot water until the tanning quality is drawn out. The liquor so obtained is used at various strengths as needed.

In the newer method the tan liquor is displaced by a solution of potassium dichromate, which produces its results with much less expenditure of time.

When the hides or skins are ready for the tanning process, they are put into a revolving drum, known as a "pinwheel," or into a pit in which are revolving paddles, with a dilute solution of potassium dichromate or sodium dichromate, acidified with hydrochloric or sulphuric acid. If the pin-wheel is employed, it is revolved for seven hours or longer; after which time the liquor is drawn off and replaced by an acidified solution of sodium thiosulphate or bisulphite, and then the revolution is continued several hours longer. If the pit is used, the skins are removed to another drum containing the second solution, and kept at rest or overturned for a like period.

In removing the skins from the pinwheel or vat, and in handling them after treatment with lime for the loosening of the hair, the hands and arms of the workmen are seriously injured, becoming raw if not protected by rubber gloves; even with gloves it is difficult to prevent injury, and in some establishments the workmen are relieved by the substitution of a single-bath process, in which the liquor is less harmful to the skin.

The hides are then removed from the pits, washed and brushed, followed by slow drying in the air. When partly dried, they are placed in a pile and covered until heating is induced. They are then dampened and rolled with brass rollers to give the leather solidity. Sole leather is oiled but little. Weight is increased by adding glucose and salt.

Various rapid processes of tanning have been devised in which the hides are suspended in strong liquors or are tanned in large revolving drums. It is claimed that this hastens the process, but the product has been criticized as lacking substance or being brittle.

Chrome tannage has been chiefly developed in this country during the last twenty years and is now in general use. It consists in throwing an insoluble chromium hydroxide or oxide on the fibers of a skin which has been impregnated with a soluble chromium salt—potassium dichromate. Other salts like basic chromium chloride, chromium chromate, and chromic alum are also used. The hydrochloric or sulphuric acid acts by setting free chromic acid.

Tanning Process
Showing the vats, the unhairing and liming processes.

Tanning Process
Showing the rotating drums.

After several hours, the skin shows a uniform yellow when cut through its thickest part. It is then drained and the skin worked in a solution of sodium bisulphite and mineral acid (to free sulphur dioxide). The chromic acid is absorbed by the fiber and later reduced by sulphur dioxide.

In the making of chrome black leather each tanner has his own method. Contrary to the general belief, there are many different methods of chrome tannage. No two tanneries employ just the same process.

Tanners of chrome leather seek to produce leather suitable for the particular demands made upon it by the peculiarities or characteristics of the varying seasons. Summer shoes require a cool, light leather; at other times a heavier tannage is essential, with some call for a practically waterproof product.

All leathers, whether vegetable or chrome-tanned, must be "fat liquored." That is to say, a certain amount of fatty material must be put into the skin in order that it may be mellow, workable, and serviceable. This is very essential in producing calf leather. Fat liquors usually contain oil and soap, which have been boiled in water and made into a thin liquor. The leather is put into a drum with the hot fat liquor; the drum is set in motion, and as it revolves the leather tumbles about in the drum and absorbs the oil and soap from the water. It is the fat liquor that makes the leather soft and strong.

Leather used in shoes is divided into two classes: sole leather and upper leather.

Sole leather is a heavy, solid, stiff leather and may be bent without cracking. It is the foundation of the shoe, and therefore should be of the best material. The hides of bulls and oxen yield the best leather for this purpose.

The hide that is tanned for sole leather is soaked for several days in a weak solution (which is gradually made stronger) of oak or hemlock tan made from the bark. Oak-tanned hide is preferred and may be known by its light color. A chemical change takes place in the fiber of the hide. This is a high-grade tannage, and is distinguished principally by its fine fibers and close, compact texture.

Oak sole leather, by reason of its tough character, and its close, fibrous texture, resists water and will wear well down before cracking. It is by many considered better than other leather for flexible-sole shoes, requiring waterproof qualities.

Sole leather is divided into three classes according to the tanning—oak, hemlock, and union.

Oak tanning is as follows: the hides are hung in pits containing weak or nearly spent liquors from a previous tanning, and agitated so as to take up tannin evenly. Strong liquor would harden the surface so as to prevent thorough penetration into the interior of the hides. After ten or twelve days, the hides are taken out and laid away in fresh tan and stronger liquor. This process is repeated as often as necessary for eight to ten months. At the end of this time the hide has absorbed all of the tannin which it will take up.

Hemlock tanning is similar to the oak tanning in process. The hemlock tan is a red shade. Hemlock produces a very hard and inflexible leather. It is modified by use of bleaching materials which are applied to the leather after being tanned. It is sold in sides without being trimmed, while the oak is sad in backs, with belly and head trimmed off.

Hemlock leather is used extensively and almost principally for men's and boys' stiff-soled, heavy shoes, where no flexibility is required or expected. Its principal desirable quality is its resistance to trituration, or being ground to a powder, and its use in men's and boys' pegged, nailed, or standard screw shoes is not in any way objectionable to the wearer. In fact, for this class of shoes, it is probably the best leather that can be used. But when hemlock is used in men's and boys' Goodyear welt shoes, where a flexible bottom is expected and required, it generally does not give good results. It cannot satisfactorily resist the constant flexing to which it is subjected, and after the sole is worn half through, the constant bending causes it to crack cross wise. On this account it becomes like a sieve, and has no power of resistance in water, and therefore it is not at all suited to flexible-bottomed shoes.

In "union-tanned" hides, both oak and hemlock are used and the result is a compromise in both color and quality. This tan was first used about fifty years ago. Twenty-five years ago the union leather tanners began to experiment with bleaching materials to avoid the use of oak bark, which was becoming scarce and high priced, and eventually developed a system of tanning union leather with hemlock or kindred tanning agents, excluding oak. The red color and the hard texture were modified by bleaching the leather to the desired color and texture. This produces leather which has not the fine, close tannage of genuine oak leather and at the same time lacks the compact, hard character of hemlock leather. Union leather produced in this manner is a sort of mongrel or hybrid leather, being neither oak nor hemlock. On account of its economy in cutting qualities, however, it is largely used in the manufacture of medium-priced shoes where a certain degree of flexibility is required in the sole. This is particularly true of women's shoes.

Union leather is sold largely in backs and trimmed the same as oak, though not so closely.

Sole leather is also made nowadays by tanning the hides by the chrome or chemical process. This leather is very durable and pliable and is used on athletic and sporting shoes. It has a light green color and is much lighter in weight than the oak or hemlock leather.

Many kinds of hide are used for sole leather. This country does not produce nearly enough hides for the demand, and great quantities are imported from abroad, although most of the imported hides come from South America. Imported hides are divided into two general classes, dry hides and green-salted hides.

Dry hides are of two kinds, the dry "flint," which are dried carefully after being taken from the animal and cured without salt. These generally make good leather, although if sunburnt, the leather is not strong. "Dry-salted hides" are salted and cured to a dry state. Dry hides of both kinds are used for hemlock leather only, although all hemlock leather is not made from dry hides.

Green-salted hides are used in making oak-tanned leather as well as hemlock, and those used by United States tanners come largely from domestic points; but there is a variable amount imported each year from abroad, principally from Europe and South America. Green-salted hides are of two general classes, those branded and those free of brands.

Cow and steer hides of the branded type are used by tanners of oak and union leather. Those not branded are used more largely for belting and upholstering leathers, a small part finding their way into hemlock leather.

Sole leather remnants, strictly speaking, include such a wide variety of items that it is difficult to cover them all. Few people, however, realize the big range of usefulness of this class of stock. While not exactly a by-product, remnants are often classed as such. Under the class of sole leather remnants are included sole leather offal, such as heads, bellies, shoulders, shanks, shins, men's heeling, men's half heeling, men's and women's three- and four-piece heeling, etc. Stock that cannot be used in the shoe business goes into the chemical and fertilizer trade, among other outlets. By a special acid process of burning this stock, ammonia is derived from it, which goes into fertilizer; and another by-product is sulphuric acid for the chemical trade. The amount of ammonia obtained is small, being about seven per cent of ammonia to a ton of sole leather scrap. This is mixed with fertilizer and sold mostly in the Southern States, and to a small extent in the West, there being a law in many of the Western States against the use of fertilizer made from leather products, on account of its low grade.

Sole Leather Offal
Showing bellies, shoulders, etc.

In the disposition of offal, heads are used for taps, top lifts, and under lifts. Shoulders are used for outsoles and inner soles, while bellies are used for medium to heavy taps and counters. Lightweight bellies and shanks are utilized for making box toes and counters.

Shanks are also used for taps and under lifts. This stock is solid and substantial and well suited for these purposes. The bellies, being flexible, are the best part of the hide obtainable for inner soles.

In cutting out soles, the manufacturer accumulates a considerable quantity of solid or center pieces, which are used for small top lifts, also for "Cuban" tops, thereby using up the bulk of the small heavy scrap that ordinarily would be sold for pieced heeling. There is also a demand for similar stock from the hardware trade, where it is used for making mallet and tool handles, also for wagon and carriage washers. Large quantities of men's and women's heeling and half heeling go to England, where it is cut up by heel manufacturers into lifts and sectional lifts for the English trade; there being a shortage of this class of offal there.

The shoe manufacturer, after cutting his soles and taps, is obliged to skive them to get the particular iron he needs. This leaves what is known as a "flesh sole shape," also a "tap shape." These skivings are pasted together by another class of trade and again used for inner soling and taps in the cheaper grades of shoes. Smaller skivings, or waste, after sorting out the sole and tap shapes, are sold to the leather board trade. This eventually comes back to the shoe trade in the shape of leather board and is cut into heel lifts. The waste after cutting heel lifts is again resold to the leather board trade and makes another round trip to the shoe manufacturer. This illustration, as well as many others in the leather remnant business, demonstrates the scientific principle that nothing is ever entirely lost. In regard to pieced heel lifts, these are made in either two, three, or four sections. This refers to what are known as sectional heel lifts. Scrap leather is also used for shanking for the European trade.

Soles and taps, known as rejects, that is, those thrown out by the high-grade trade, are sold to manufacturers of cheaper lines. A shoe manufacturer cutting his own soles and buying sole leather in sides, after sorting out the soles suited to his own requirements, will sell what he cannot use to remnant dealers, who in turn re-sell them to shoe manufacturers requiring that particular class of stock. The scrap leather, or remnant dealer, thus forms a useful link in the chain of distribution, furnishing a market where shoe and leather manufacturers may dispose of their surplus products to best advantage, and providing a source of supply for buyers who wish any particular article to suit their individual needs.

Upper or dressed leather is made from kips or large calfskins. It is tanned and finished like all other forms of leather by variations of the foregoing process. Thick hides are often split thin by machinery, and the parts retained and finished separately. The parts of the leather from the hair side are most valuable and are called "grain" leather; the inner parts or "flesh splits" are made into a variety of different kinds of leather by waxing, oiling, and polishing.

It is finished by scouring with brushes and then rubbed with a piece of glass, which removes creases and wrinkles and stretches the leather. Then it is stuffed with a mixture of oil, soap, and tallow, which is worked into it by rolling. Various finishes are given to leather, such as seal grain, buff, glove grain, oil grain, satin calf, russet, plain shoe, etc.

Upper leathers are blacked by rubbing with a mixture of lampblack and oil or tallow, or with a solution of copperas and logwood.

No tanning process, no matter how good or thorough, can make firm, serviceable, wear-resisting leather out of all portions of any hide, because nature made some parts of every hide porous, spongy, and lacking in fibrous strength.

Calfskins used by tanners are of several classes. American calfskins, taken off in the United States and Canada, are usually sold green pelted. Farmers raise only a small fraction of the calves born. Each cow must produce a calf in order to insure a maximum flow of milk. Most of the farmers keep cows to produce milk, hence they sell the young calves for veal and use their skins for high-grade calf leather.

In European countries farmers fatten their calves before selling them in order to get a higher price for the veal. The skin is not so valuable for leather as the skin from younger calves, and it is used for lower-value leathers.

Calfskin is not split. A heavier weight skin might be. It is shaved to a uniform thickness.

Calf leather is divided into the following classes, depending upon the finish of the leather:—

Boarded calf (made in both chrome and bark tannage).

Wax calf, finished on the flesh side with a waxy, hard surface.

Box calf is a proprietary name. It is boarded—rubbed with a board to raise the grain. It is known by minute, squarelike lines.

Mat calf is a dull-finished calfskin, used more in topping.

Suede calf is finished on flesh side. Most makes of suede calf are chromed, although there are some vegetable.

Storm calf is a heavy skin, finished for winter wear. Considerable oil is used in finishing.

French calf is finished on flesh side.

Dry hides are obtained from Buenos Ayres, where the cattle are raised on the plains. This city exports a large quantity of hides, dry, salted, and cured by smoking. The hides of cows generally yield inferior grain leather; but South American cow-hides may be worked for light sole leather.

Calves' hides are thinner, but when well tanned, curried, and dressed, they yield a very soft and supple leather for boots and shoes. They are finished with wax and oil on the flesh side, and can also be finished on the hair (grain of skin).

Calves' skin (green salted).

Paris City calfskins. These are obtained in three grades—light, medium, and heavy.

Light grades run from four to five, or seven to eight pounds; medium grades run from seven to nine pounds; heavy grades run from nine to twelve pounds.

Patent leather may be made from colt, calf, or kid skin. Coltskin is the skin of young horses, or split skins of mature horses.

Patent colt and kid are used for the most part in the medium fine grades, and patent side (cowhide) is used in the medium and cheaper grades. Chrome tanned are used entirely in the manufacture of patent leather.

Patent leather, as it appears in shoes, may be described either as varnished leather, coltskin, or kid, and sometimes the French use calfskin. The process is largely a secret one, although there is no longer any patent on the principle of the same. It is made by shaving the skins on the flesh side or hair side to a uniform thickness. Then it is de-greased to put the skin in condition to receive the finish and protect the same from peeling off. Successive coats of liquid black varnish are applied, the first coats being dried and rubbed down, so as to work the liquid thoroughly into the fibers of the leather. The last coat is applied with a brush, and baked to from one hundred and twenty to one hundred and forty degrees Fahrenheit for thirty-six hours and then allowed to dry in direct sunlight for from six to ten hours, which seems to be essential to remove the sticky feeling. Various ingredients are used in making the different varnishes, the first coating consisting of naphtha, wood alcohol, amyl acetate, etc. The black varnishes consist of linseed oil and various other mixtures, heated in iron kettles. The final coating is a naphtha preparation resembling japanning material. The hide is stretched on a frame during the varnishing operations.

It is almost impossible to tell the difference in quality of shiny leather by appearance, although in general the leather on which the grain shows through the varnish will prove more serviceable than that on which the finish is so thick as to hide the grain. Great care must be exercised in resewing patent leather shoes that have been exposed during the cold weather, as the cold has a tendency to freeze the finish. Patent leather, like all varnished coatings, is liable to crack. No one can guarantee it not to do so. The kid patent leather is more elastic and porous than other kinds. The serious objection to the use of patent leather for a shoe is its air-tightness. This makes it both unhygienic and uncomfortable. The kid patent leather is the only patent leather that has not this objection.

Kid is a term applied to shoe leather made from the skins of mature goats. The skin of the young goat or kid is made into the thin, flexible leather used for kid gloves, which is too delicate for general use in shoes. The goats from which come the supply of leather used in this country for women's and children's fine shoes are not the common, domesticated kind known in this country, but are wild goats or allied species partially domesticated, and are found in the hill regions of India, the mountains of Europe, portions of South America, etc.

There are about sixty-eight recognized kinds of goatskins that are imported from all over the world. The Brazilian, Buenos Ayres, Andean, Mexican, French, Russian, Indian, and Chinese are a few of the many kinds that are known as such. Each particular species of goat hide possesses its own peculiarities of texture. The thickness and grain differ according to the environment in which the animal has been raised. It is peculiar that those raised in cold climates do not have as thick skins as those raised in warmer climates, for the long, thick hair apparently takes the strength.

We may wonder where all the skins come from that are made up into glazed kid, mat kid, and suede, at the rate of several thousand dozen every day. The great proportion of the skins are *goatskins*. These are almost all imported from abroad, where the animals are slaughtered and disposed of much the same as we dispose of beef and veal here. Sheepskins and carbarettas, the hides of animals that are a cross between sheep and goats, are also used.

The finer grades of kid and goatskins which are tanned in large quantities in New England, come from the Far East.

In China there are two great ports from which skins are shipped, Tientsin and Shanghai. Back in the interior, starting from a point about twelve hundred miles from the sea, collectors make their rounds twice a year.

The breeder of goats kills his flock just before the collector is due, skins the animals on the hillside, preserves the meat for food, and with the kidskins, which have been partly dried, wrapped in a bundle carried upon the back, or upon a pack animal, the breeder makes his way to the station. It may be that there are a half hundred breeders awaiting the coming of the collector and he pays them the market price for the skins.

Whenever the collector has a sufficient supply to make it profitable to ship, he bales the skins and then sends them over the thousand mile journey along the river to the seaport. From Tientsin or Shanghai they are taken by tramp steamers, which reach Eastern ports by way of the Suez Canal, and on the trip the steamers make several ports, so that it is from six to ten weeks before the skins reach America.

Another method of importing is to have the raw material shipped across the Pacific and then transferred to a railroad, but the difference in cost to the manufacturer is so great that it is unprofitable.

The China goatskins are rated as among the finest in the world and when tanned they make the highest-grade shoe.

Then there are mocha skins, which come from Tripoli, Arabia, and Northern Africa. In those places the method of collection is practically the same as in China.

The two best-known grades are the Hodieda and the Benghazi. They derive their designations from the exporting cities. Hodieda is located in the southwestern part of Arabia on the Red Sea, while Benghazi is in Barca, one of the provinces of Tripoli.

Other goatskins are produced in India and Russia, and millions of skins are exported annually from Bombay, Madras, and Calcutta. These skins are not brought direct to America, but are transshipped at Marseilles or London.

The jobbers in Europe or India occupy rather a unique position, for according to their practice it is almost impossible for them to suffer financial losses in dealing with an American tanner. The latter, when he wishes to arrange for his year's supply of raw material, negotiates with an agent in Boston, with whom he signs a contract for so many skins. Then it is necessary for the tanner to either purchase with money equal to the face value or secure by loans letters of credit from Boston banking houses which have European connections.

Before the skins are exported, the jobber has his money from the European banking concerns and the bills of lading are forwarded to the Boston bankers, who turn them over to the tanners, and, when the occasion requires, obtain from the tanners what is known as a deed of trust.

All goatskins are tanned by the same chrome tanning process, whether the finish is to be glazed or mat. The proportions of chemicals vary according to the texture of the skin, and according to the grain.

The process of tanning is quicker than the tanning of heavier hides, and all varieties of tannage are used, the chrome methods having come into very general use. There are many kinds of finish given, such as glazed, dull, mat, patent, etc. One quality that distinguishes goat leather, the "kid" of shoemaking, is the fact that the fibers of the skin are interlaced and interlocked in all directions. The finished skins as they come from the tannery, by whatever process they may be put through, are sorted for size and quality, a number of grades being made. Instead of ripping straight through, like a piece of cloth, or splitting apart in layers, as sheepskin will do when made into leather, the kid holds together firmly in all directions.

Glazed kid is colored after it is tanned by submerging it in the color, a very important process. The glossy surface is obtained by "striking" or burnishing on the grain side. It is made in black and colors, particularly tan, and is known by about as many names as there are manufacturers of it.

Glazed kid is used in the uppers of shoes, making a fine, soft shoe that is particularly comfortable in warm weather, and is said to prevent cold feet in winter, owing to unrestricted circulation.

Mat kid is a soft, dull black kid, the softness being the result of treatment with beeswax or olive oil. It is finished on the grain side the same as glazed kid, and is used principally for shoe toppings. It is very similar in appearance to mat calf and is often used in preference to it, as it is of much lighter weight, and about as strong.

Suede kid is not tanned, but is subjected to a feeding process in an egg solution, called "tawing," to make it soft and pliable. The skin is stretched and the color is applied by "brushing" (with a brush). The color does not permeate the skin, but is merely on the surface. Suedes are made from carbarettas and split sheepskins. Suedes are used very extensively in making slippers, and come in a great variety of colors.

A castor kid is a Persian lambskin finished the same as a suede, and is used in making very soft, fine-appearing leather—like glove leather. The skin is of such a very light weight that it has to be "backed up" before being made into shoes.

Fancy leathers are used extensively for toppings of shoes having patent leather vamps. Facings are selected from fancy leathers to make the inside of a shoe attractive and to increase its wearing quality. Leathers of dull or glazed finish are used in typical shoe colors.

Miscellaneous kinds of kids are as follows:—

    A. Kangaroo
    B. Buckskin
    C. Sheepskin
    D. Chamois
    E. Cordovan
    F. Splits
        a. Seal Grain
        b. Buff
        c. Oil Grain
        d. Satin Calf
    G. Enamel
    H. Sides

Kangaroo is the skin of the animal of that name.
Buckskin is the skin of certain deer.
Sheepskin is the skin of the familiar domestic sheep.

Chamois is the skin of the animal of that name and by courtesy the specially treated skins of certain domestic animals.

It is a simple matter to recognize a kid-skin among the various kinds of upper leather, because of its very light weight and pliability.

During the winter, leather, in drying, is apt to become frozen, especially where no well-equipped drying loft is provided. Such leather becomes weak and limp if thawed out too rapidly. In freezing, the water in the hides which have been hung up to dry is forced out and stretches apart the hide fiber. The wetter the hides, therefore, the more demoralized they will be by the frost. The treatment of rushing the frozen leather into a warm room is inadvisable; the best method is to allow the hides to hang as they were and to tightly close all openings to the outside air. In case this is impossible, it is best to place the leather in a heap, in a room where the temperature will not fall below the freezing point, and to cover it with a cloth. In case the leather rolls up, it should be moistened before the roll becomes greater than is customary; it will thus become firmer throughout. Some upper leather and especially sheepskins for lining purposes are aided by freezing, since the leather becomes white and plump and also is of a bright color, though the durability is somewhat lessened.

The popularity of white leather for shoes is increasing wonderfully. There is good reason for this. The modern white shoes have a stylish and fashionable appearance which has won the hearts of women of all ages and conditions, and when they want a thing, there is always alertness in supplying it. The new love for white shoes is interesting to trace. Years ago white leather for shoes was made principally from deerskins. But this leather, while attractive when new, would stretch soon after being worn, and take on a yellowish tinge. Besides, the price of such shoes was very high, and it is not surprising that they became supplanted by the cheaper, but attractive and useful, white canvas shoes, which became quick sellers during the season.

It is greatly to the credit of our tanners that they have been able to perfect and put on the market a white leather for shoes which answers all requirements satisfactorily. This leather is made from cow hides; the white color will not fade nor turn yellow, and best of all, the leather can be easily cleaned and made to look good as new. Another advantage is that such leathers can be used in shoes that sell at popular prices.

There are many common, commercial grades of upper leather.

Willow calf is a fine, soft, chrome tannage of calfskin. It is sold in three colors, light tan, ox blood, and olive-brown. The distinguishing features of this leather are its durability and the fact that it always keeps soft and pliable. It is adapted to the highest quality of men's and women's shoes.

Box calf is a storm-calf leather of highest quality. It is a waterproof, chrome tannage in a medium tan color, with a dull finish. This is the best leather obtainable for rough, outdoor wear, walking shoes, hunting boots, etc. It is also adapted to men's and women's very fine footwear. There is a growing demand for this kind of shoe. In the uppers of the best storm shoes you will always find box calf.

Royal kid is a black chrome calfskin, dull finished with a smooth, natural grain of fine texture, soft and pliable. It is used for vamps and whole shoes of the highest grades for men and women, and is a very popular material for the fall and winter shoe. The desirable qualities of fine calf leather are making the demand for it grow faster than the supply of raw material increases.

Tan royal is a tan color, chrome calf leather, smooth finish, fine grain, excellent cutting qualities, uniform, of medium rich tan shades. Tan calf leather is very attractive and the tan shoe is now a staple product.

Cadet kid is a bright black, smooth-finished, chrome calfskin for men's and women's fine shoes. This tannage and finish give a remarkable cutting value. The stability of this stock is entirely unique and makes the finished shoe stand up, keeping its much desired shape through the different tests of manufacturing. It is said to be the best calfskin, by the best judges, the shoe manufacturers.

Bronko patent is distinguished for its fine, coltskin-effect grain. It has a rich and lustrous black patent finish. The results obtained from bronko patent in its workings through the shoe factory and its wearing qualities afterward have never been equaled. Bronko is one of the finest results of the development of chrome patent leather.

Cadet kid side is a chrome side leather that closely imitates the calfskin, called cadet kid. It has a bright, lustrous finish, and a remarkably fine grain. It is surprisingly like fine calf leather in appearance.

Cadet calf sides are similar to cadet kid sides with the exception of a boarded finish. This is another black chrome, side leather which comes very near to a calfskin.

Mat royal chrome side is a special finish, closely resembling calf, used for the tops of men's and women's medium fine shoes.

Black hawk patent is a well-tanned, well-finished patent leather for medium-priced women's shoes and for tipping.

Colored box chrome side, boarded, is a substitute for willow calf.

Black box chrome side, boarded, is a substitute for box calf in medium fine shoes.

Kangaroo kid side is a back-tanned, dull, smooth, black leather nearly like calf, used in the tops of men's shoes, and men's and women's whole shoes.

Waterproof black is a high quality leather of great durability for men's and boys' heavy shoes. Waterproof brown is similar to waterproof black, except in color, and is a leather made for hard service.

Amhide black is a soft, dry, high-grade tannage for lightweight, comfortable, sporting, work, and hard-wear shoes.

Amhide russet is like black amhide in everything but color.

Hercules storm chrome is a leather distinguished for its fine grain and good appearance of medium heavy weight.

Boris is a heavy-weight, soft, water-proof leather for men's medium quality shoes. It is finished in three colors and black.

Zulu is a medium-priced leather, which makes a very fine wearing heavy shoe. It is made in two colors and black.

Bison is a colored or black-finished leather, of a high grade, very comfortable and durable.

Ottawa is of two colors and black finished, and is suitable for high quality, heavy, rough shoes.

Sheboygan calf is a heavily stuffed, soft, water-proof leather. It is of two colors and black.

Dongola calf is a black leather used for durable, medium-priced, heavy, outdoor shoes.

Belt knife splits are sold in several tannages and finishes of the most improved manufacture. These splits are sorted in all weights. Uniform selection is maintained, and the quality in every way is of the highest order.

Oxford calf union splits is one of the highest grades of grain-finished, union splits. It has an extremely soft and fine appearance.

Cambridge calf union splits have a most careful and high-grade finish, but somewhat firmer than Oxford calf.

Flesh splits are sold in two tannages. These are the highest-grade flesh splits that it is possible to make, and they are a long distance ahead of the ordinary, flesh splits, their improved finish making them a modern and largely used substitute for satin.

Ottawa black and russet splits include a variety of printed splits, used for shoes in combination with grain leather and for whole shoes. They are selected in many weights.

Flexible splits for Goodyear, gem, McKay inner soles, is leather that offers the greatest advantages to large and small buyers. It is the product of six different tanneries, assorted in all the usual weights. Great care is taken in the manufacture of these splits to adapt them perfectly to the shoe manufacturer's needs.

Flexible bends are used by manufacturers of Goodyear welt shoes requiring a straight Goodyear or gem inner sole. They find these bends of great advantage on account of the small amount of waste, the strength and desirability of stock. They are made in six tannages.

Chrome flexible splits for inner soles furnish a very strong and durable leather for inner soles, taps, and outer soles.

Ooze gusset splits, colored, give a very low-priced leather suitable for gussets, bellows tongues for high-cut boots, also for the quarter-linings of Oxfords.

Ooze vamp splits, black and colored, are strong, durable, low-priced leathers suitable for cheap work shoes where water-proof qualities are not required.

Chrome-tanned embossed splits, colored, are made in a great variety of patterns for cheap shoes and other work where leather is required. They are durable and low priced.

## Leather for Belting

A native steer about four years old, killed in the month of October, affords the best example of a good hide for belting manufacture, that is, for the transmission of power from pulley to pulley. At this age and at this season the steer is in prime condition.

On account of the great and enormous strain put upon belting, and the necessity for its running true upon the pulley, it should be of the highest grade possible, combining great strength to prevent stretching, and evenness of grain to insure long wear; therefore only hides of selected steers are serviceable, and these in turn are rejected when they contain any blemishes or cuts or other imperfections. After a hide is accepted for belting purposes, it is subjected to a generous trimming, the head, neck, legs, and belly being cut away, leaving only a small and compact section embracing from two to two and a quarter feet on each side of the backbone and extending about six feet along the same from the tail forward. This is the portion of the hide where the fibers are closely and firmly knit together, and where the vitality is the greatest, due to the close proximity of the network of nerves radiating from each side of the spine to all parts of the hide.

The hides of the bull and cow of every breed are inferior for belting purposes to that of the steer. The hide of the bull is coarse and hard, with the neck very heavy and full of wrinkles, causing a variation in the thickness and run of the grain of the leather. The hide of the cow is thin, does not run uniform in thickness, being heavier on the hips than at the shoulder, and is lacking in the firmness necessary in good belting. The sharp angles of the hip bones of a cow also tend to form pockets in the hide.

After the hide has been trimmed, it is subject to the process of "currying." All membranes or particles of flesh adhering to the hide are removed by a machine which shaves the membrane, etc., off, with lightning rapidity. The leather is then washed and scoured by machine, which removes all dirt still adhering to the hide. After the leather is thoroughly cleaned and while in a damp state, it is placed upon the table, and greases, composed of pure animal oil, are worked into the leather on both the grain and the flesh side with brushes. This is carried on in the cold state. It is then put into a large revolving wheel containing water heated to a high degree, which causes the leather to swell and pores to open. The leather is then taken out and put into another wheel containing heavy mineral oil and heated several degrees greater than the water, and tumbled about in the wheel until the heavy oil fills the distended pores and fibers. After this, the leather is allowed to dry.

The hides are allowed to remain for several months in the tan liquor until the green hide is changed into leather.

After the hide has been changed into leather, it is stretched. To properly stretch the leather for belting purposes, it must first be cut so as to remove that part which shows the markings of the backbone of the steer.

Leather is stretched by placing it in clamps, every part of the piece getting the same pull. (The leather is put into the clamps while damp, as damp leather will give the greatest amount of stretching.)

When the stretching process is completed and the leather has thoroughly dried in stretching clamps, it is released. These pieces of leather are quite dry, very firm, and not very pliable. The leather is now moistened in order that it shall be more pliable as it passes through the finishing processes. After the water has soaked into the leather (called sammied), it becomes very soft. It is then subjected to a roller under heavy pressure to take all the unevenness out of the hide. It is next thoroughly dried, causing the fibers to shrink; then again moistened and put through a polishing machine, which acts on the same principle as the rolling jack.

The sides and centers are now put through a cutting machine, which reduces the leather to strips of different sizes.

Belts are put together by cementing the parts. Belt cement is a most powerful adhesive. It actually governs the strength of the belt, as the belt is as strong as the weakest part of the joint.

## Rawhide Products

Rawhide is used for a great many purposes. After the side of leather has been trimmed of the portions that cannot be used, it is sold to the lace maker. He measures the same in a machine.

The trimmings from the side of the hide may be used for a mallet head or other tools made of leather. The most common products of the strong section of rawhide strings are shoe strings, belting laces, and parts of harnesses. It is also made into leather shoe strings that are used in the logging camps.

When the hide is selected for the rawhide purposes, it is first passed to a dehairing machine, where all the hair is removed. It is then fleshed; that is, all loose membrane and any flesh that may have adhered to the hide are removed from the flesh side. The rawhide is then placed in a special bath for the purpose of opening the pores, before the oils and greases are added to it. After this bath, it is dried thoroughly in a hot box and then put into wheels which mill the greases into the hide.

The hide, which is made hard by this drying process, is put through breakers, where it is thoroughly worked into soft and pliable form.

The hide is next passed to the setting-out machine, which finishes all forms of leather—by condensing and strengthening the fibers. Special oils are applied to both the grain and flesh side of the hide. It is finished by hand and cut into laces. This hand finishing is usually done in order to reject all parts that are not perfect.

Haired leather is tanned by acid—a quicker method. The hide is split into sides and tanned with the belly stock on them, which is used for car straps; cow bell straps, trunk straps, and riding bridles.

### The By-products of a Leather Belting Factory

There are a great many by-products in a leather belting factory, all of which are used. The finest strips are used for whip lashes, small pieces are used for the French heel, and the extremely small pieces are used in leather mats.

The by-product from the belting bull, which is about fifty per cent, is used for shoe leather and leather straps. There is considerable leather taken from the belting bull for certain harness work. The belly is thick and porous though not tough, and is used for halters, cow bridles, and other parts of harness where the strain is not great.

### Round Belt Making

Round belt is made from the best belting, but while the strain on round belting is not severe, the leather must be soft and pliable. It is selected from regular stock of native steer hide.

## Properties of Tanned Leather

Leather that has been tanned is made up of a great many little bundles of fibers. The coarser and stronger fibers are on the inside, and the very fine and smoothly laid fibers are on the outside. These fibers are so intertwined and so elastic that when the leather bends these bundles play on one another. On account of the smoothness of the surface it may be polished, and beautiful finishes and effects obtained on the leather.

The elasticity of leather (which is due to the elasticity of its fibers) allows it to stretch to a great extent. The tendency to return to its original position is very strong at the beginning, but grows weaker if the strain is continued at any one point. Of course, in stretching the leather, there is always a corresponding drawing in another part of the shoe, which gives it a worn and baggy appearance.

When shoes are removed from the feet, they are oftentimes damp, due to perspiration. The stretched or strained fibers are apt to shrink and return to their original position. In order to avoid this, it is necessary to place shoe-trees in them.

When the linings of shoes are exposed to friction and excretion of perspiration from the feet of some people, they deteriorate. This is due to the fact that the acids of perspiration (acetic, formic and butyric acids) have become so concentrated that they act on the fibers of the leather. These acids exert a burning effect, causing the fibers to lose their elasticity so that they no longer play on one another, but become fastened to each other. The result is that they become hard, and any attempt to bend the leather tears them apart; and once the union of fibers is destroyed it cannot be repaired.

In order to keep the fibers in such a condition (soft and flexible), they should be lubricated often (twice a week) with a liquid followed by a wax paste, usually called shoe dressing. When a brush or a piece of cloth is rubbed over the surface of leather containing the shoe lubricants (shoe polish), it produces a smooth surface called a "shine."

Compounds which shine without friction produced by brush or cloth should not be used, as they are simply varnishes and one coat on top of the other destroys the leather.

## Substitutes for Leather

In olden times our fathers and mothers used handmade shoes, and wore them till they had passed their period of usefulness. At that time the consumption did not equal the production of leather. Knowledge of conditions in the great western countries to-day will show that many of the big cattle-raising sections, once famed for their cattle, have been taken up by homesteaders and are now producing grain instead of cattle. But since the appearance of the machine-made shoe, different styles of shoes are placed on the market at different seasons, to correspond to the change of style of clothing, and shoes are often discarded before they are worn out. We have not been able thus far to utilize cast-off leather as the shoddy mill uses wool and silk, etc. The result is that the consumption of leather is above the production, therefore substitutes must be used.

In shoe materials there is at present an astonishing diversity and variety. Every known leather is used from kid to cowhide, and textile fabrics have developed rapidly, especially in the making of women's and children's shoes. The satins, velvets, serges, and other fabrics that are used in the manufacture of shoes must be firm and well woven, and are usually supplied with a backing of firm, canvas-like fabric, to give strength.

As to wearing quality the old saying, "There is nothing like leather," still holds good; but people do not buy shoes for their wearing qualities alone in these days. Style and intrinsic beauty are considered, and; have a cash value just as in any other article of apparel.

Each fabric is made of two sets of thread-like yarn woven at right angles to each other. They are called the warp and filling (weft). The warp is composed of yarn running the longest way of the fabric, and filling runs the short way of the fabric. Since the warp is the body of the cloth, it is its strongest part and all fabric in shoes should be placed warpwise across the foot of the wearer, so as to be able to resist the great strain.

Various attempts have been made for legislation to prohibit the treating of leather by chemicals or the use of substances to increase its weight. Complaints have been made by a number of shoe manufacturers that the excessive use of glucose (a form of sugar) in sole leather has resulted in injuring the leather and fabrics composing the uppers of shoes.

Representatives of large leather firms claim that the methods of tanning sole leather have radically changed during the last few years, and that the small quantity of glucose and epsom salts that is used to-day in finishing sole leather is absolutely necessary to its value, and is in no sense an adulterant or weighting material. Shoe manufacturers, on the other hand, claim that in some cases larger amounts of glucose, salt, etc., have been added to the soft leather from the belly of the animal, in order to give it the desired stiffness. On account of the high price of leather, various attempts have been made to find a substitute for it. Most of these substitutes consist of strong cloth treated with some drying oil like linseed, the oil having previously been mixed with other solid substances.

A prize of five thousand francs has been awarded to a Belgian inventor, Louis Gevaert, for his unusually superior artificial leather. The process consists in the more or less intimate impregnation of stout cloth with tannic albuminoid substances. Shoes made of this are said to possess not only the resistance and elasticity of natural leather, but its durability of wear. Moreover, they are much cheaper, costing, including manufacture, only four francs (about eighty cents) and being sold at about six francs per pair.

# CHAPTER FOUR.

## THE ANATOMY OF THE FOOT

VERY few people, even among those engaged in the shoe industry, know much of the anatomy of the foot. Yet it is evident that they ought to know something about it in order to furnish the foot with a proper covering.

The first thing that strikes a person on looking at the human foot is its large proportion of bone. On pressing its top surface and that of its inner side, the amount of flesh will be found to be very small, indeed. The same is true of the inner and outer ankle. The extreme back of the ankle has scarcely any flesh covering. The most fleshy portions of the foot are its outer side, the base of the heel and the ball of the big toe.

The reason for this disposition of flesh is to protect or cover those parts of the foot that support the body by coming in contact with the ground. They act as pads and lessen the concussion. The abundance of flesh on the outer side of the foot is to protect or act as a shield against danger. The inside of the foot is not exposed as much as the outside.

The foot is divided into three parts, the toes, the waist and instep, and the heel and ankle. The largest bone of the foot is the heel bone (called calcaneum). It is the bone that projects backward from the principal joint and forms the main portion of the heel. When a person is flat-footed, this bone is thrust farther backward than nature intended to have it. The connection between it and the tarsal bones is lost.

The top bone of the foot is the astragalus, and it forms the main joint upon which the process of walking depends. This bone has a smooth, circular, upper surface that connects it with the main bone of the lower leg. It is absolutely necessary that this bone should be in perfect harmony (relation) with the others in order to insure comfort and health. If the arches of the foot are forced out of position, up or down or sidewise, this joint is not permitted to do its work normally.

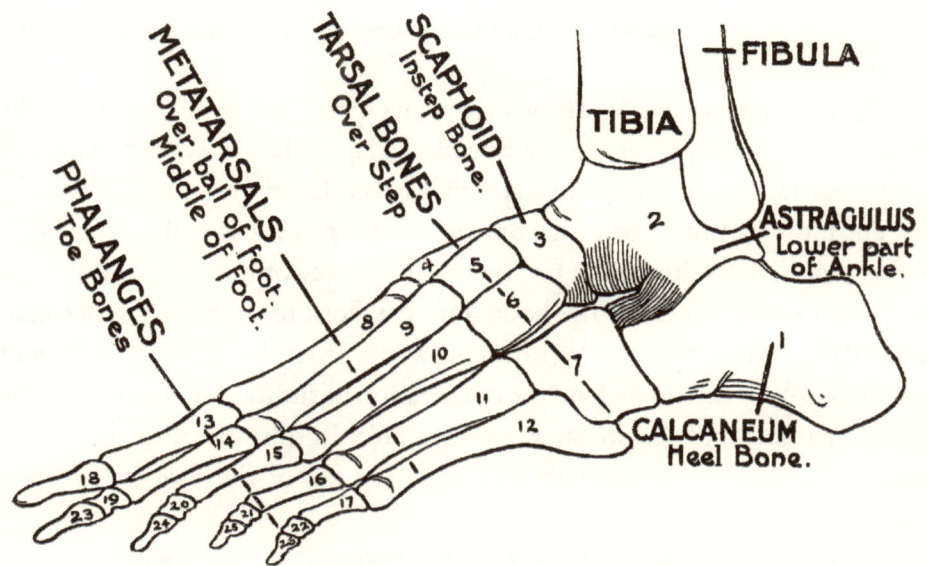

The Bones and Joints of the Human Foot.

Rheumatism is a frequent evil of an injured joint. Hence the necessity of absolutely normal action, unhampered by ill-fitting shoes.

The principal arch of the instep is called the cuneiform or tarsal bone. Persons are troubled with defective insteps to quite an extent. Misshapen joints at this point due to shoes that do not fit and consequently disarrange and throw out of position the delicate, natural structure, work great havoc with the comfort of the foot. Nine joints cluster at this point.

The bones of the toes are called the metatarsal bones and phalanges. There can be no doubt that nature intended mankind to walk in his bare feet, and in that event the phalanges of the foot would occupy a much more important part than is now the case as a result of modern civilization. There are nineteen bones in the foot, and the disturbance of one or more of these will serve to upset the entire foot by throwing out of relationship the general unit of work devolving upon the whole number of joints and bones. Each joint has its accompaniment of muscles, and each lack of alignment of bones and joints provokes discord and lack of harmony in the muscular action.

Muscles are attached to bones, and by their contraction or extension the bones are moved. Very few movements are effected by means of a single muscle. The muscles of the foot in nearly all cases are in combination, and are so complex in their action that the best surgeons find it difficult to describe them satisfactorily.

The chief characteristics of the foot are its spring and elasticity. While the foot has wonderful powers of resistance and adaptability, it is the shoemaker's duty not to strain the same, but to provide for each action.

The most sensitive part or the one part that is most susceptible of injury is the big toe. This is due to the fact that the tendency of the foot in walking is to travel toward the toe of the boot, and in a word to press into rather than shun danger. The shoemaker provides for this, first, by allowing sufficient length of sole to extend beyond the termination of the toe, and second, by the fit of the upper and the preparation of the sole. In this way, if the toe of the shoe strikes against a hard substance, the big toe of the foot will remain untouched.

Seventy-five per cent of the people have more or less trouble with their feet. Some of these troubles are caused by the manufacturer putting on the market shoes whose lines look handsome and attractive to the eye, but are lacking in any other good features. Shoes that fit properly should have plenty of room from the large toe joint to the end of the toes, and also should have plenty of tread, especially at this point.

A mere glance at our bare foot will show conclusively that pointed-toe boots are false in the theory of design. The toes of a foot when off duty touch each other gently. When they are called on to assist us in walking or in supporting our body, they spread out—although not to any great extent. This, then, being the action, no sensible maker of boots and shoes would attempt to restrain them. Box or puff-toe shoes allow the greatest freedom.

The pointed-toe shoes, which join the vamp to the upper immediately over the big toe joint, exceedingly high heels, and thick waist shoes are not for the best interests of the foot.

The evils of ill-fitting shoes are corns, bunions, and calluses.

Corns are mainly due to pressure and friction. When the layers of skin become hardened, they form a corn, which is merely a growth of dead skin that has become hard in the center. This hardened spot acts like a foreign body to the inflamed parts.

A hard corn is formed more by friction than pressure. It is produced by the constant rubbing of a tight or small shoe against the projecting parts of some prominent bony part, as the last joints on the third, fourth, and little toe. When this action continues, it produces inflammation. Rest—as relieving the feet of the friction—decreases this inflammation, leaving a layer of hardened flesh. Renewed action reproduces the same effects, leaving behind a second layer of hardened flesh. This continued action and reaction brings on a callus, rising above the surface of the skin. This increases from its base. An ordinary hard corn may be removed by scraping up the callous skin around its border, and prying out carefully with a knife. Soft corns are chiefly the result of pressure or friction. These corns are soft and spongy elevations on the parts of the skin subjected to pressure. Soft corns are mostly found on the inner side of the smaller toes. Those on the surface of joints by mechanical action will become hard.

The blood corn is excessively painful. It is the result of an ordinary corn forcibly displacing the blood vessels surrounding it, and causing them to rest upon its surface.

The bunion is an inflammatory swelling generally to be found on the big toe joint. The chief cause of bunions is known to be the wearing of boots or shoes of insufficient length. The foot, meeting with resistance in front and behind, is robbed of its natural actions, the result being that the big toe is forced upward, and subjected to continuous friction and pressure. The wearing of narrow-toe boots that prevent the outward expansion of the toe is another cause.

The comparisons of quantities are often called ratios. The ratios of the different parts of the foot to the height are different in the infant from that of the adult period. Between these two periods the ratios are constantly changing.

There are two series of shoe sizes on the market; the smallest size of shoe for infants (size No. 1) is, or was originally, four inches long; each added full size indicates an increase in length of one third of an inch (sizes 1 to 5). Children's sizes run in two series, 5 to 8, and 8 to 11; then they branch out into youths' and misses'; both running 11½, 12, 12½,13 ,13½ and back again to 1, 1½, 2, etc., in a series of sizes that run up into men's and women's. Boys' shoes run from 2½ to 5½; men's from 6 to 11 in regular runs. Larger sizes usually are made upon special orders. Some few manufacturers go to 12. Women's sizes run from 2½ to 9. Some manufacturers do not go above size 8. The rate of sizes is sometimes varied from by manufacturers of special lines of shoes. A man's No. 8 shoe would be nearly eleven inches long. These measurements originated in England and are not now absolute.

A system of French sizes is used which consists of a cipher system of markings to indicate the sizes as well as widths so that the real size may not be known to the customer.

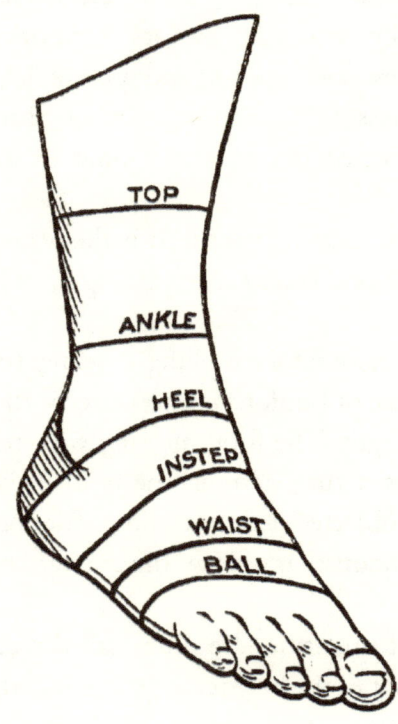

The Different Parts of the Foot and Ankle.

All feet are not alike in structure and shape. In infancy the foot is broad at the toes, which press forward in the direction of their length. The heel is small in comparison to the width of the toes, and also short in length, due to the undeveloped bones. But during growth, the thickness above the heel bones disappears, and the heel itself becomes thicker and assumes the beauty of perfection at maturity. This development is due to the growth of bones which must be well exercised and properly cared for during this period. The various parts of the feet and legs do not mature at the same rate—those at the upper part of the body increase at a greater rate than the lower parts. Thighs develop first, next the upper part of the legs, and lastly the feet.

The adult foot, when properly formed, is straight from heel to toe on the inner side, and is wider across the joints than one inch or so farther back. The manner of walking has a considerable bearing on the character and development of the foot.

There are many sorts of feet, which are due to a number of causes, such as habits, climate, occupation, locality, etc. As a general rule we may divide the feet into four classes: Bony feet—those with very little flesh upon them; hard feet—those that have plenty of flesh, but which are almost as hard as a stone; fat feet—plump, with plenty of flesh, but having little shape; spongy feet those that seem to have no bones in them, usually found in the female sex.

The characteristics of a foot are common with the body to which it is connected. Some people have a strong, bony frame, with strong, firm muscles, prominent bones and muscles, and a flesh that is hard. The feet of this type of person are usually long, bony, and arched, with a well-developed big toe joint. The heel measurements are large in proportion. A soft foot is prevalent among the Scotch. The feet of a person who is delicately shaped, with a small frame and thin, small, tapering muscles, are usually thin and finely formed, giving evidence of quickness. This kind of a foot in a man has a tendency to develop a flat foot.

A person with a form inclined to plumpness, full of exercise and activity, and a good circulation, has a well-developed foot. The heel is round and fairly prominent, although there are no special bony prominences. On the other hand a person with a body of general roundness, but with tissues and muscles flabby, and a languid blood circulation, has feet that are short, soft, and flabby.

We will allow that these four different kinds of feet all measure a 4 size and D in width. One would naturally think that the same size shoe would fit them all, but this is not so. This size shoe will only fit one and that is the bony foot. The hard feet require a C½ width; the fat feet require a C width, and the sponge feet require a B width.

The same last may, and often will possess a slight variation in some manner or other. The fitter of feet must know the stock, each pair, and be on intimate terms with the peculiarities of each last and the inside lines of each pair of shoes before attempting to try them upon the feet of the customer.

Different makes of footwear are apt to be manufactured over a slightly varying system of measurements. One line of shoes made over a small measure may be longer or shorter or narrower or wider than some other line. The heel measurements require careful study for each line introduced. The peculiarities of each line must be tested by tape and measure, and the foot fitter must have a strong knowledge along these lines.

We should measure the foot by the stick if necessary, and make a note of the size and width that will be likely to prove a fit. The height of the arch must be considered, and the shape of arch curve, the shape of the instep, and the general contour of the foot. A normal foot will show about a half-inch arch. The average foot will carry from an inch to an inch and a quarter heel, without putting a strain on any of the joints of the foot. Some feet vary from this by a wide margin. A foot is a trifle longer in walking than in repose. Allowance should be made, in using the measuring stick, over what the foot actually draws on the stick. In men's shoes the allowance should be from two to two and one half sizes.

When a one-legged man buys a shoe, the dealer sends to the factory a shoe to match the one left remaining. In these days of the use of machinery in every process of their manufacture, shoes are made with the utmost exactness and precision, and it is easily possible to mate that remaining shoe with the greatest nicety in size, style, material, and finish.

Few people have feet exactly alike; commonly the left foot is larger than the right, so that one shoe may fit a little more snugly than the other. Commonly, however, people buy shoes in regularly matched pairs, the difference in their feet, if it is noticeable to them at all, not being enough to make any other course desirable.

But there are people who buy shoes of different sizes or widths, in which case the dealer breaks two pairs for them, giving them, to fit their feet, one shoe from each. In such cases the dealer matches up the two remaining shoes, one from each of two pairs just as he would where he had broken one pair to sell one shoe to a one-legged man.

But a man does not have to be one-legged nor to have feet of uneven sizes or shapes to make him ask the dealer to break a pair of shoes for him. A man with two perfectly good feet came into the store where he was accustomed to buy and wanted one shoe. While traveling in a sleeping car, his shoes had been mixed up with others and he had received back one of his own and one of some other man's; a fact which he had not discovered until he was too far away from train and station to set things right. So he came in to buy one shoe to match his own.

# CHAPTER FIVE.

## How Shoe Styles are Made

IF you examine the shoes worn by people in a large city, you will notice different styles. Shoe styles that were called grotesque a few seasons ago are comparatively usual to-day, for the new designs in women's footwear, which manufacturers are now making, are the most varied that ever have been put on the market. Pink and green and blue are among the new colors in materials for footwear.

Some of the styles for the coming seasons are more lavish than have hitherto been seen in the women's shoe trade of America. Coronation purple velvet boots look like an extravagant color for footwear, but they are now selling. Samples of pink, green, and blue shoes, both boots and pumps, are being made up, and they will soon, be offered to buyers.

The style of the shoe is dominated by fashion. All styles are related, that is, every part of our dress is influenced by the prevailing fashion, ideas of color, fabric, or garment outline. To illustrate: when short skirts are stylish, women wear mannish shoes to harmonize with them; on the other hand, with long skirts they must have a shoe that is neat and small, hence, the short vamp. When women wear white in the summer, cool canvas shoes spring into favor; when gray and blue dress materials are to be used, a variety of tan shoes are worn to harmonize, etc.

After the style has been decided upon, it is necessary to work out an exact reproduction. An expert model maker, called a last maker, produces a last, a wooden model of the shoe. In order to do this, it is necessary to lay out certain plans or specifications for the details of the manufacturer of the shoe.

There are certain parts of all feet that have fixed measurements. To illustrate: the length of the shank, that part of the sole of the foot between the heel and ball, in every person's foot is always the same. The part of the foot back of the ball or large toe joint conforms to certain fixed measurements.

These definite measurements form a basis by which the last maker originates new styles by shortening, lengthening, widening, or narrowing the space in front of the toes, but always retaining the true and fixed measurements of the back part of the last.

When the last maker desires to produce a new style, he takes an old last and tacking pieces of leather on some parts of it (front of the toes), he builds it up and cuts off other parts. This patched-up last is taken to a special machine (lathe), where a number of duplicates are turned from a block of wood.

The "pattern maker" is the man in the factory who makes patterns, consisting of heavy pieces of cardboard bound with brass, in the shapes of the various pieces of leather required to make up the upper part of the shoe.

The pattern maker has found by experience that the top part of the shoe also conforms to certain fixed measurements, and by working in sympathy with the last maker he need only to change the front part of the vamp to bring out the latter's ideas. With these measurements as a foundation, he puts forth from time to time different style uppers, as buttons, lace, blucher, fixings, scrolls, straps, ties, pumps, etc. This is the way new style tops originate.

After the manufacturer has approved of sample patterns, the pattern maker receives an order for a certain quantity of patterns to be made over a certain last which is submitted to him. Working on the fixed top measurements and the last submitted as a basis, the pattern maker draws plans for a model pattern. The standard size of a model pattern is size 7 in men's shoes and size 4 in women's. He is also given an order for a certain number of widths; for instance, B, C, D, and E, and he draws out on paper a complete set for each width in the size 7. These four sets of model patterns are reproduced and cut out in sheet iron by hand. But from these sheets any number of iron models, and any size regular cardboard pattern can be reproduced by a machine.

Wood to be made into lasts comes to the shoe manufacturers in a rough, unchiseled form. The lasts are made of maple wood; hollow forms used by traveling salesmen and window trimmers are made of bass wood.

The making of the model of, the last is the most exacting operation in the factory. It is produced by a machine most important. The principle of this machine has been brought about by the pantograph; that is, it will turn from a rough block of wood an exact copy of the model last; or it will enlarge or reduce a duplicate of any other size or width, so, from a single model last, such as the manufacturer has decided on, any number of lasts can be made, and of any size or width. The machine itself consists of two lathes. On one is placed the model and on the other the block of wood. The model is held against a wheel by a spring. By adjusting this wheel, any desired width last can be obtained, and by adjusting a bar in front of the machine any length last can be produced from the block of wood.

Rough Unchiseled Block of Maple.

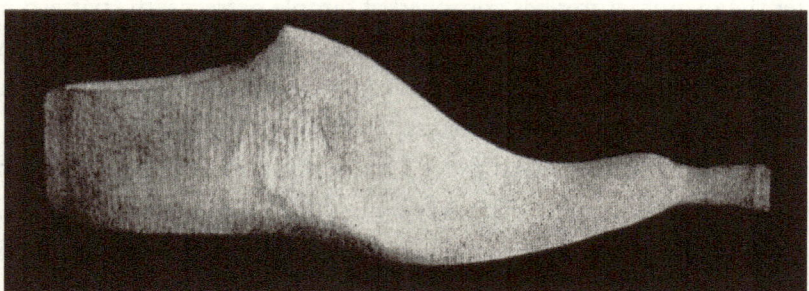

A Last after leaving Turning Lathe.

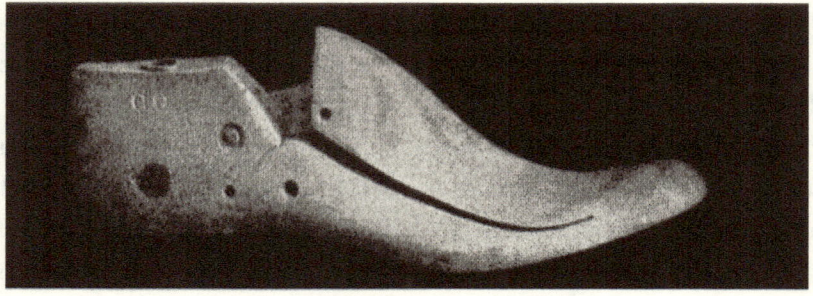

A Finish Last.

The lathe, when in motion, revolves both the last and the model, the model being pressed against the wheel, which is really a guide for the revolving knife that digs into the block of wood, and regulates the depth that the knife is allowed to cut. In this manner the model is reproduced from the block which is also regulated as to size and width by the wheel and by the bar. This machine is so accurate that a tack driven into the model to locate the center of the last is reproduced by a sort of a wooden pimple in the block of wood when finished. The model sole pattern is now tried on the half-finished last to insure accuracy.

Notice in the figures of lasts that the turning lathe has left stubs of wood on the toes and heels. These must be finished to a "templet." The templet is a measure or guide used to indicate the shape any piece of work is to assume when finished. From the heel and toe of the model, a piece of iron is shaped on an exact arc of that model, and is used on the heeler machine as a guide to form an exact copy of the heels and toes of the model. This machine works very rapidly, and by the aid of an irregular shaped, revolving knife it quickly transforms the toes and heels to the desired shape. The bottoms are again tried out on a sole pattern and the last number, the size and the width are stamped on.

We now have the last as a solid piece of maple wood and turned to the desired shape, size and width. Were it possible to insert and extract the last in this form from the half-finished shoe, no other steps would be necessary in last manufacture, but inasmuch as the leather is stretched very tightly over this last a little later, it necessitates the introduction of some method that will facilitate a quick removal of the last from the shoe. This is accomplished by cutting it in two parts and making a hinged heel. The fact that the slightest measurement changes the size of the shoe, necessitates great care in the introduction of the hinge as a part of the last, and in order to insure accuracy and uniformity in all the lasts, they are marked with templets and gigs. The hinge must be placed inside of the last.

The finished last is so constructed that it can be readily inserted or withdrawn from the shoe, and the strong hinge provides the last, when inserted, with the same rigid qualities as though it were one piece. The center of the last is indicated, as before stated, by a reproduction in the side of the last of the tack that was placed in the model. This is the mark that locates the position of all the holes, and it is done by a "gig" in the following manner:—

A gig is a piece of steel having cylinders that guide the bit of the boring machine in an exact perpendicular line. This gig, being placed on the last in the position marked by the turning machine, forms the accurate location of the bolt holes that hold the hinge.

After the hinge is placed in the last, it goes to the ironers to have the bottom put on it, if it is a McKay last, and a heel plate if it is a welt. The bottom is again tried and the plate filled up to the same. The last is then ready to go to the scouring room. In this room the last goes through three operations, first of which is ruffing. This consists of scouring with a coarse grade of quartz. This operation must be carried on so that the sole lines and insteps are not brought into contact with the quartz.

The second operation, medium grinding, is done with a fine grade of quartz, and in this operation, also, the worker keeps away from the toe. The third operation is done with a much finer-grade quartz, the operator going over the entire last. The last is now ready for polishing, and after that, for a heavy coat of shellac. It is polished and waxed on a leather wheel. Then it goes into the shipping room ready for shipment to the manufacturer.

# CHAPTER SIX.

## DEPARTMENTS OF A SHOE FACTORY— GOOD-YEAR WELT SHOES

THE principal methods of manufacturing shoes are the following:—Goodyear welt; McKay; turned; standard screw; pegged; nailed.

The simplest and the clearest way of showing how the various kinds of shoes are made is to explain the manufacture of a Goodyear welt and afterwards bring out the points in which this method of shoe-making differs from the others.

Shoes are manufactured in up-to-date factories, employing hundreds of operatives. The modern shoe factory of to-day is divided into six general departments: the sole leather department, upper leather department, stitching department, making department, finishing department, and the treeing, packing, and shipping departments.

In some sections of the country, several of these departments are often designated by other names. The stitching department is often called the fitting department; the making department, the bottoming department; and the sole leather department, the stock-fitting department. The departments are popularly termed rooms for brevity.

A shoe factory is designed so as to have a width of about fifty feet for each room, while the length is according to the number of shoes to be produced. A width of about fifty feet gives plenty of daylight and ample room in the center of each department, which is very essential in shoemaking.

Shoe factories are usually about two hundred feet long, while many are nearly four hundred feet. A few exceed four hundred feet, running as long as eight hundred feet. Some are built in the shape of hollow squares, while others have wings added, which give almost as much floor space as the original building.

A Modern Shoes Factory.

The average factory has usually four floors. The first floor, or basement, is occupied by the sole leather department. The next floor above includes the treeing, finishing, packing, and shipping departments, and also the office. The third floor is devoted entirely to the making or bottoming department. The top floor is divided so that the cutting and stitching departments have each half a floor.

There are several exceedingly large factories in this country that find it advantageous to divide the factory into more departments, as, for example, the cutting room is divided so that the linings and trimmings are cut in a separate department. The skiving may also be done in a separate room. The making room will be divided so that the lasting is set off as a separate department on account of the many workmen and machines employed. In the same way there will be a division of work so that the packing and shipping will be set apart from the treeing. Then, again, in the sole leather room, the making of heels as well as the fitting of the bottom stock may become independent departments.

The system of making women's shoes is practically the same as that of men's except that in a great many factories the method of preparing the bottom stock is somewhat different. Most manufacturers of women's shoes do not cut sole leather, but buy outsoles, insoles, counters, and heels, all cut or prepared. These soles are in blocked form and large enough so that they can be cut or rounded by the manufacturers to fit their lasts. The counters, when bought, are all ready to put in the uppers, while the heels are ready to put on the shoes. Whenever a manufacturer of women's shoes cuts his sole leather, he has the same system as that in the men's factories.

In women's factories where sole leather is not cut, they do not have a complete sole leather department. Instead, they have what is called a stock-fitting department. There are independent cut sole houses, etc., in the trade, which supply the soles to manufacturers. The same system of buying supplies also applies to many other parts of the shoe, as in the top lift, half sole, welt, rand, etc. In the upper leather department, manufacturers of both men's and women's shoes often buy trimmings and other parts of the upper all prepared.

A large proportion of the men's shoe manufacturers are now buying heels all built, while fully nine tenths buy counters all molded. The soles and other parts that are needed for a shoe are put up in different qualities and grades, and a manufacturer can buy any grade of sole he wants, so that it is considered an advantage to buy some parts, instead of cutting them. In a side of sole leather there are twenty-five or more different qualities and grades of soles, and very few manufacturers, especially in the women's trade, can use all of these. The greater variety of shoes a manufacturer turns out, the more advantageous it is for him to cut his own sole leather, and prepare all parts in his own factory.

In this country the number of factories in the shoe trade appears to be growing less and the average factory larger each year. It is estimated that there are at present something like fifteen hundred factories in all. These range from the smallest product up to the largest. The average factory may be said to produce about twelve hundred pairs of shoes per day. Many turn out five thousand pairs daily, while a few manufacturers turn out ten thousand or more pairs. Several manufacturers and firms have half a dozen or more factories and have a total output of between twenty thousand and thirty thousand pairs of shoes a day. There is no such thing as a trust or monopoly of any kind in this trade, and there never has been up to the present time.

In all factories and all classes of work, the "case" has always been of such a number of pairs that it can be divided by twelve in every instance. A case can be twelve, twenty-four, thirty-six, forty-eight, sixty, or seventy-two pairs, and in children's work it is often sixty and seventy-two pairs. All cases of these numbers are regular cases, whereas any other number would be out of the ordinary. Of course, a case of shoes may contain any number of pairs, but the numbers given above have always been used in regular work.

Cases of shoes may differ, but every pair of shoes in any one case must be made exactly alike. All shoes are made in cases, except in the matter of custom work or single-pair orders or samples. In making men's heavy shoes, or working shoes, the regular case was formerly sixty pairs or thirty-six pairs, but the tendency has been of late to have a standard case of twenty-four pairs. In the men's fine trade the regular case is twenty-four pairs, while in the women's it is thirty-six pairs. Long boots for men have always been made in twelve-pair cases.

Goods are sold by the samples, sent out with the traveling salesman. As fast as he receives an order, he sends it to the main office. Here the orders are subdivided and sent to the factories making the goods. For example, an order for seventy-five dozen men's shoes of a certain style received by the main office from the traveling salesman would be sent to the factory in the form of a typewritten order, covering the general description and sizes written out in the proper form, for each case is made according to the specifications on the tags that are made out in the office. These tags specify the sole, heel, upper, kind and quality, how stitched, the last to be used, how bottomed finished, treed, and packed. Everything is marked plainly on the tags so that a buyer can have any shoe made just as he wants it.

This order would be sent from the factory office to the cutting room, where a clerk would make out twenty-five long tickets.

Twenty-five are made because the shoes go through the factory in lots of twenty-four pairs, each lot being called a job and when finished making a case of shoes. The long ticket is made in duplicate form, and is perforated so it may be tied to a lot of shoes. Both parts of the tickets are made out to contain the various operations with the specifications as to detail. The lower part is sent to the stock or sole leather room, while the top part remains with the uppers which are cut in the cutting room. While each part of the ticket is sent by a different route through the factory, they finally meet in the form of finished shoes.

In addition to the long ticket already described, two other tickets are made out, the top ticket and the trimming ticket. The top ticket is sent to the leather bins of the factory, where the sorter knows by experience exactly the amount of leather required to cut the order, being careful to see that it is all of uniform quality and free from blemishes. He rolls the leather in a bundle, attaches the ticket and sends it to the cutter.

In the cutting room there are three classes of cutters; cutter of trimmings, who cuts lace stays, top facings, back stays, tongues, etc.; outside cutter, who cuts quarters, vamps, tops, tips, etc.; and the lining cutter, who cuts cloth linings.

Skins of leather are received in the shoe factory in different shapes. Some are perfect, others have blemishes or imperfect spots. The skins that are to be used for upper stock are carefully graded by two or three men, as to quality of leather and weight. This is necessary in order to be sure that a lot of shoes made for a certain dealer will be uniform. On account of the leather coming in different shapes, some skins perfect, others having imperfect spots, the cutter must place his patterns in such a way that certain parts of the shoe will use up all the perfect parts, and others, less important, will be composed of the weaker parts of the skin. This explains why you sometimes find the inside top part of a shoe made of flanky leather, while the vamp is made of a better grade.

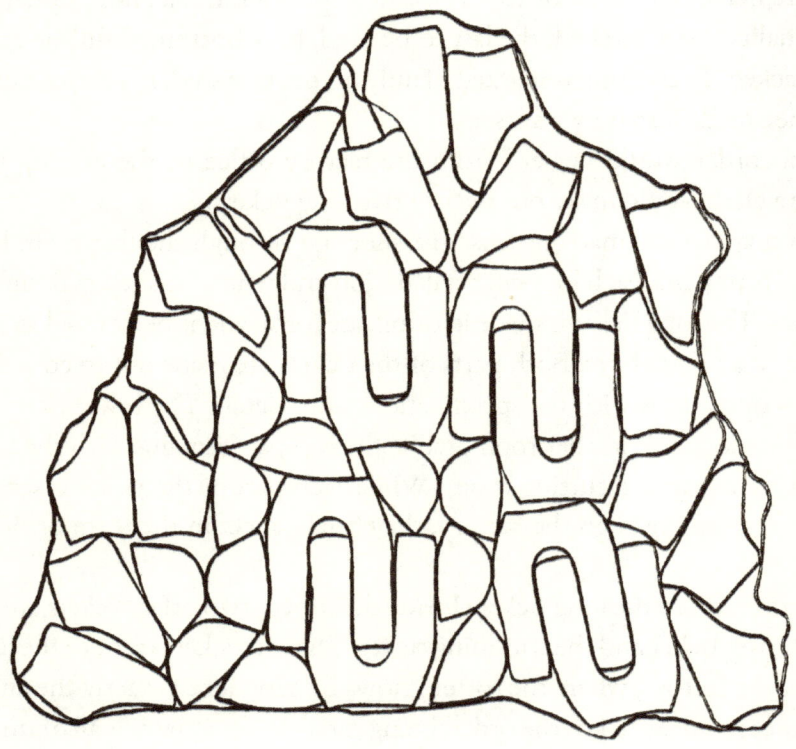

A Nine and One-Half Foot Skin divided to best advantage before Cutting.

There is a pattern for each and every size shoe, and each piece of leather is cut out separately on a block of wood. Nothing is wasted. In order to make each cutter as efficient as possible, the cutters are divided, so as to have a different cutter for each grade of leather. In this way they become better judges of leather.

The lining cutters use patterns and knives on drilling. The facing is cut out with a knife and pattern. The side stays and the tongue are cut out by dies.

After the leather has been cut into the desired shape, uppers, vamps, toe pieces, back stays, lace stays, etc., cutting at times ten pieces, and for some styles of shoes as many as fourteen pieces, the cutters take care to keep the parts for the same shoe together, matching and marking them so that eventually all will meet again in the shoe.

Machines are used now on almost every operation, and annually several new machines make their appearance. The cutting of uppers up to four or five years ago was performed by an operator cutting the leather by running the knife along the side of the pattern. Now they are using a cutting machine and dies to cut uppers in nearly all factories. This cutting machine is called the "clicking machine," and it is considered quite a labor saver in a department where it was the universal opinion that machines never could be used.

It is impossible to give a list of all the operations performed and have it complete. But a good general idea of the system can be given and the name and meaning of the main operations in the several departments. It should be kept in mind that the methods in rooms differ, and that hardly any two factories put a shoe through in exactly the same manner. The general system and plan is the same everywhere and the machines are the same in all factories, but the details and minor operations are so numerous that there is plenty of scope for them to vary.

The function of the clicking machine is to cut the upper leather into the desired shapes required. It consists of an iron frame, with a cutting board on the top of it. Above this is a large beam which can be swung to the right or left of any portion of the board. The skin to be cut, which may be of any kind, is placed on the board and a die of the design or shape of the leather desired is placed on it. The handle of the swinging beam is taken by the operator and moved over the die; then by pressure of the handle the beam is brought downward, pressing the die through the leather. As soon as this is done, the beam automatically returns to its full height.

These dies are made in different designs and sizes to meet the different sizes and designs in the upper of the shoe. One die for each design and size. They mark the vamps for the location of the toe cap and blucher foxings as well as the size by means of nicks in the edge of the piece cut. The dies are about three-quarters of an inch in height and so light that they do not mar the most delicate leather.

After the outside cutter has cut the skin into pieces to make up the shoe, these are tied up in separate bundles, that is, the twenty-four of tips in one bundle, twenty-four pairs of vamps in another. These are turned over to girls who stencil the sizes on the edge and match them, that is, see that each upper is exactly like the mate.

After the different parts have been cut by the operator of the clicking machine or by hand, the edges of the upper leather, which shows in the finished shoe, must be thinned down (skived) by a "skiving machine" to a beveled edge. This is done in order that the edges of the leather that are to show in the completed shoe may be folded to give a more finished appearance. The machines are operated by girls; each one an expert on one particular piece.

The order number and size of shoe are stamped on the top lining of each shoe. After all linings have been prepared, according to the data given on the instruction card attached to parts of the shoe, the parts are sent to the stitching department, where the stitchers on a multitude of machines stitch all the different parts together very rapidly and accurately.

The toe caps are then given a series of ornamental perforations along the edge. This is done by either "power tip press," or a "perforating machine." The first consists of a series of dies placed in a machine by which the leather is perforated according to the designs desired. Each series of dies represents a different design.

Cutting the Leather by Means of Pattern and Knife.

The perforating machine resembles a sewing machine, but instead of a series of dies, the one in this machine is made of single or combination dies which make one or more holes on each downward movement. The machine feeds automatically and does the work very accurately. The cutting tool is kept from becoming dull by pressing against a band of paper. Ornamentation on other parts of the shoes, such as the edges of vamps, etc., is made by this machine.

Before going to the stitching room, every bundle is examined by sorters. The sorters are divided and subdivided; that is, one man always sorts tips, another vamps, etc. They examine each piece for imperfection, and if any is found, the piece is thrown out and a new one put in. The last operation is the assembling of pieces. Here each job of twenty-four pairs is brought together and securely tied and numbered.

Goodyear Stitching.
A machine that sews around the edge of the welt and joins it to the sole exactly at the heel.

This stitching department is one in which female labor is generally employed, although in late years more men are being used to operate machines, especially on vamping or other heavy parts. In some parts of the country it is called the fitting room. The work of the department consists of stitching the different parts of the upper together, so that it is ready to put on the last. The terms used mean in most cases stitching the part named to the rest of the upper. There are very many operations in the department, several of which are named below, together with their meaning.

The bundles of pieces which have come from the cutting room are placed on the table, where they are subdivided into three parts, the linings, the tops, the vamps and the tips.

The linings for the tops of the shoes are pasted together (with the back strap and top bands), care being taken to join them at the marks made for that purpose. After being dried, they go into the hands of the machine operators, where they are joined together by a stitching machine, and the edges, etc., trimmed. The sewing machines used are very similar to an ordinary home sewing machine, with the exception that they are much larger and stronger.

The lining is finished. The next step is to join the lining to the piece of leather making up the outside of the same shape, called the top. The top receives the eyelets by a machine placed in proper position. The top and lining can be put together by sewing them face to face. The top is inspected and all threads clipped off.

After the shoe uppers have been properly stitched together, the eyelets are placed on by a "duplex eyeletting machine, which eyelets both sides of the shoe at one time. The top of the eyelets are solid black knobs, so as not to wear brassy, while the bottom (which clinches inside the shoe) called the barrel, is of nickel. This finishes the shoe upper.

The vamp, tongues, and tip are then put together. The edges of the vamps, quarters, tips, etc., are covered with a cement made of rubber and naphtha, which is kept in small bowls on the benches in front of employees. Several grades of cements are used. The cemented parts are allowed to dry, and the edges are then turned over by "pressing machines," which gives a finished appearance. The shoe is put together by stitching the vamp to the quarters. This work is done by both men and women, and is work which demands much care.

Stock Fitting Room.

Where all bottom stock is prepared after being cut.

In stitching men's uppers, the system varies in various factories as much as it does on women's. Here are some of the operations, which will give an idea how men's uppers go through.

Extension or toe piece sewed to vamp.
Leather box stitched on.
Tip stitched to vamp.
Vamp seamed up back.
Top folded around edge.
Top seamed up.
Eyelet row stitched up and down.
Lining seamed up.
Side facing put on lining.
Top facing put on lining.
Lining and outside pasted together.
Under trimming.
Eyeletting.
Hooking.
Vamping.

The upper is complete when it leaves the stitching room and is all ready to be put on the last. While the upper is being prepared, the soles, insoles, counters, and heels are made in other departments.

When the foreman of this department has received the tags with the data necessary for the preparation of outsoles, insoles, counters, toe boxes, and heels, they are sent to the stock room, where these parts are kept.

The soles are roughly cut out by means of dies, pressing down through the leather, in "dieing out machines." Before the soles are cut, the leather is dipped in water and sufficiently dampened. After they are cut out, they are made to conform to the exact shape by rounding them in a machine called the "rounding machine." The roughly died out piece of leather is held between clamps, one of which is the exact pattern of the sole. The machine works a little knife that darts around this pattern, cutting the sole exactly to conform. The outsole is now passed to a heavy rolling machine, where it is pressed by tons of pressure between heavy rolls. This takes the place of the hammering which the old time shoemaker gave his leather to bring the fibers very closely together, so as to increase its wear.

Counters and toe boxes (stiffening which is placed between the heel and toe cap and the vamp of shoe) are prepared in the same room with the heels. After they are made, they are sent to the making or bottoming room, where the shoe upper is awaiting them. As the counter is an important feature in the life of a shoe, much depends upon the quality of leather that goes into it.

The sole is next fed to a "splitting machine," which reduces it to an absolutely even thickness. The insole is made of lighter leather than the outsole, but has the same thickness and is cut out in the same way one at a time. The sizes are stamped on them and they are sorted.

If you examine a Goodyear welt shoe, you will notice no stitches in sight, the seam being fastened to an under portion of the insole. The durability of the shoe relies, to a great extent, on the quality and strength of the insole.

Welting.

The smooth-appearing insole of a welt shoe must be either pasted in or fastened underneath in some manner. This fastening is accomplished by passing the insole through a very small machine called a Goodyear channeler, which makes two incisions at one operation. It cuts a little slit along the edge of the insole, extending about one-half inch toward its center.

The upper part of insole made by the slit on the edge is turned up on a lip turning machine so that it extends out at right angles from the insole. In other words, the channel is opened up and laid back, forming a ridge around the outer edge of the sole. This forms a lip or shoulder, against which the welt is sewed. In this way the thread used in sewing cannot be seen in the finished shoe. The cut made on the surface serves as guide for the operator of the welt sewing machine when the shoe reaches him.

The inner and outer soles as well as the uppers are now brought into the lasting or gang room. The first part of lasting is called "assembling," which means that many parts are brought together, such as upper, counter, insole, box toe, and last. The counter is placed in the upper, between lining and vamp, while the box toe is shellacked and put in the toe of the upper (provided it has not been stitched in the stitching room). The operator first tacks the inner sole on to a wooden last.

Lasting.

There are very many different styles of lasts, and in cutting uppers a different pattern must be used for each style. Then the upper is placed in position on the last, and it is ready to be pulled and stretched to the wood and take its required shape. This is accomplished by placing the shoes on the "pulling over machine," where the shoe uppers are correctly placed on the last by the pincers of a machine holding the leather at different points securely against the wood of the last. By the movements of levers the shoe uppers are adjusted correctly. Then the pincers draw the leather securely around the last and at the same time two tacks on each side and at the toe are driven in part way, to hold the uppers securely.

It is now placed on the "hand method lasting machine, where the leather is drawn tightly around the last. Before this operation, it is dipped in water to preserve its shape when formed and that it may be more easily formed by the machine. At each pull of the pincers a small tack, driven automatically part way in, holds the edge of the upper exactly in place, so that every part of the upper has been stretched in all directions equally. A special machine by means of a series of "wipers" is used to last the toe and heel. After the leather has been brought smoothly around the toe, it is held there by a little tape fastened on each side of the toe, which is held securely in place by the surplus leather, crimpled in at this point. The surplus leather crimpled in at the heel is forced smoothly down against the insole and held there by tacks driven by an ingenious hand tool. In all these lasting operations the tacks are only driven in part way, so they may afterwards be withdrawn and leave the inside perfectly smooth, except at the heel of the shoe, where they are driven into the iron heel of the last and clinched.

After these operations, the surplus leather at the toe and sides of the shoe is removed by the "upper trimming machine," which cuts it away by means of a little knife and leaves it very smooth and even. A small hammer operating in connection with the knife pounds the leather on the same parts. A pounding machine hammers the leather and counter around the heel so that the stiff position conforms exactly to the last.

After the "lasted" shoe has been trimmed and pounded down to the shape of the last, it is turned over to the tack setter, who pulls out all the tacks except a few, called draft tacks. The insole is then wet to make it pliable, and is turned over to a very experienced operator, called the "in-seamer," who is to sew the welt on.

The shoe is now ready to receive a narrow strip of prepared leather, that is sewed after it is wet to make it pliable, along the edge of the shoe, beginning where the heel is placed and ending at the same spot on the opposite edge. This is called the welt, and is sewed from the inside lip of the insole, so that the curved needle passes through the lip, the upper, and the welt, uniting all three securely and allowing the welt to protrude beyond the edge of the shoe. The thread is very stout linen, and is passed through a pan of hot wax before being looped into chain stitch that holds the shoe together.

The nature of the stitch is a chain—two rows of threads on the outside that loop with the single thread in the inside lip of the insole. When the welt is finally sewed on, and the shoe put down on the bench, it looks like an ordinary shoe resting on a wide flange of leather. This flange is the welt, and to it the heavy outer sole is to be sewed fast. Should a single stitch break in this operation, it is passed to a cobbler, who repairs it by hand.

Edge Trimming.

Before the outer sole is put on, the edges of the uppers must be trimmed along the seam that holds the welt. A slip of steel called steel shank is laid along the insole where the hollow of the foot is, and a piece of leather board laid over this to give the necessary stiffness and prevent the shoe from doubling up. As the welt has left a hollow space along the ball of the foot, it is necessary to fill this up, either with a piece of leather, tanned felt, or other filler. Felt is not waterproof, and leather squeaks, hence a mixture of ground cork and rubber cement is used. This is heated and spread on the sole, and run over a hot roller until the bottom of the shoe is perfectly smooth and even. The shoes are placed on a rack and are ready for the outsole.

Rough Rounding.

Sole fastening is performed by a number of operations, in which a score or more of separate machines are used. The sole layers smear a rubber cement over this welt with a "cementing machine," after the outsole has been soaked in water to make it pliable, and then place it on the shoe and tack a single nail in the heel. The "sole laying machine," through great pressure, cements the sole on and fits it to every curve of the last. Then the sole is trimmed by a "rough rounding machine," which trims the soles to the shape of the last. This machine also channels the outer sole at the same time, which is necessary for the next operation. The "channel opening machine" now turns up the lips of the channel and the sole is ready to be stitched to the welt.

The outsole is now sewed by a waxed thread to the welt, by an "outsole lock stitch machine," which is similar to a welt sewing machine. The stitch is finer and extends from the slit (channel) to the upper side of the welt, where it shows after the shoe has been finished.

It unites the sole and welt with a tightly drawn lock stitch of remarkable strength. It sews through an inch of leather as easily as a woman would sew through a piece of cloth. The stitches are made through the welt and outer sole, the seam running in the channel of the outsole.

The inside of the slit in which this stitch has just been made is now coated with cement by means of a brush. The channel lip is forced back to its original position after the cement has dried, by a rapidly revolving wheel of a "channel laying machine." In this way the stitches are hidden.

Welt shoes are stitched on in three different ways: "channeled," which, when finished, leaves an invisible stitch on the bottom of the sole; "regular stitched aloft," showing the stitches on both sides; and "fudge stitched," in which the seam is sunk down in a groove, being almost invisible from the welt side.

Every stitch must be of such a nature that it is independent of the one next to it, so that should one stitch break, the others will not work loose. This is accomplished by running the threads through a pan of hot wax just before entering the leather, which causes the waxed thread to solidify, becoming, as it were, a part of the leather.

Notice should be taken of the difference between the way the outsole is stitched and the inner sole is stitched to the upper. In place of three threads in the chain stitch "that holds the welt to the upper and insole" there are but two here—an upper and a lower one. The upper thread extends only part way down, where it loops, twists, and locks into the lower thread. This is the reason why you can wear a welt sole clear through without its pulling loose.

Shoes that are stitched aloft go through the same operations as the channel-stitched shoes, with the exception that the rounding machine contrivance of cutting is eliminated.

Shoes that are to be fudge stitched are sent through the same machine as the regular stitched aloft, but an additional little knife point on the arm of the Goodyear stitcher digs a channel in the welt so that the stitches on that side are sunk into the leather.

The outsole is nailed at the heel after the stitching on the "loose nailing machine," which drives the nails through the outsole and insole and clinches against the steel plate of the last. The machine drives separate nails fed from the hopper of any desired size or length, at the rate of three hundred and fifty per minute.

The edge of the outsole around the heel is now trimmed to conform exactly to the shape of the heel on the "heel seat pounding machine."

The stitches of the regular stitched shoes are separated by a series of indentations, giving the shoe that corrugated effect which adds so much to the appearance of the shoe. In the fudge-stitched work the stitches "are entirely covered up by the indentations.

Leveling.

Then a leveling machine, called the "automatic sole leveling machine," with a pressure of about two and a half tons to each of the concave rollers, comes into play. The rolls move automatically back and forth and from side to side, doing the work that the shoemaker used to do on his lap with a hammer and stone, but doing it better and more quickly. It practically levels off the bottom of the soles.

An automatic gauge regulates exactly the distance from the edge of the last, and by the use of this machine the operator is enabled to make a sole conform to that of all others of a similar design and size.

Heels are formed by cementing different lifts of leather. A machine called a "heel cutter" shapes out the lifts. The heel is then placed under pressure, giving it exact form and greatly increasing its wear.

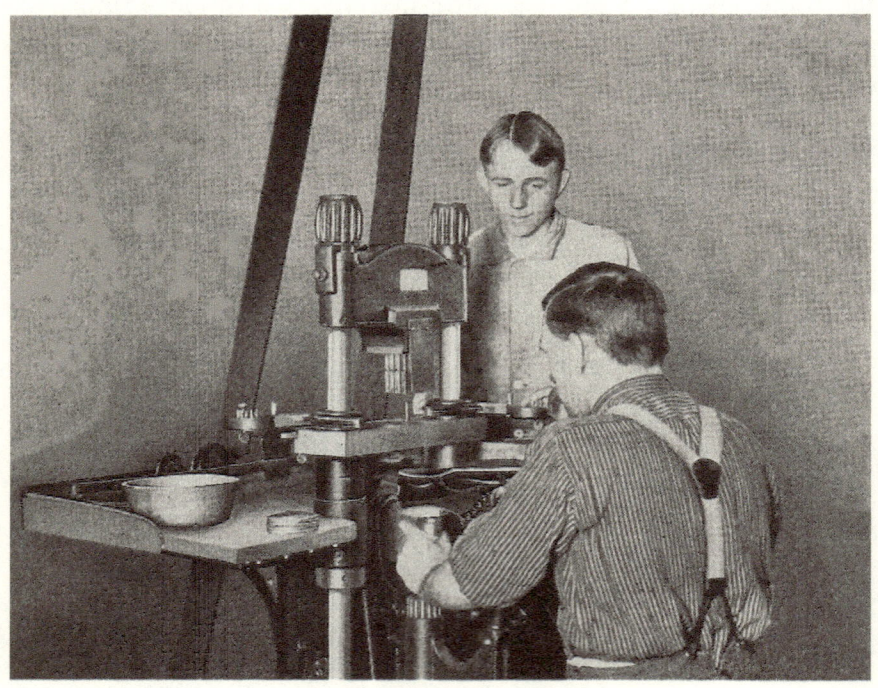

Heeling.

In speaking of the ends and sides of a heel, the part that rests on the ground is spoken of as the top, and the first piece is called the top lift. The part that is fastened to the shoe is spoken of as the bottom, while the side nearest the toes is called the breast. The wedge is a flat, heel-shaped piece or lift of leather that is skived to a thin edge at the breast. Being thicker at the back, it tips the heel forward. Wedges are made from thin strips of waste leather, or from sheets of leather board, and are cut out with a hollow die. The gouges are cut in the sole leather room from scraps, and are a regular heel lift, having a horseshoe-shaped piece of leather with an opening at the breast.

The sole leather, insoles, counters, and heels, in the stock fitting department are "got out" by being cut into shape by a machine die.

The heel is now trimmed of all rough and surplus portions of leathers to the exact size of top lift. A blower attached to the machine removes all scraps, etc.

The breast of the heel, which faces the forepart of the shoe, is trimmed evenly across and with the desired slant by means of a peculiar-shaped knife which extends over the sole at shank. The edges of the heel are now scoured by revolving rolls with molded sandpaper to make perfectly smooth. Blowers attached to the machine remove all dust.

There are several types of machines for fastening the heel to the shoe, all very rapid in operation. One of the latest is that which feeds the nails, and which is operated by a man and boy, who together turn off a great quantity of work.

Sole Scouring.

The nails are left protruding slightly above the heel so as to retain the top lift, which is now placed in position by the same operator on the same machine. It is pressed down over the heads of the nails securing it in position. The small brass or steel nails which protect and ornament the heel are now driven in by the "universal slugging machine." This machine cuts the slugs from a coil of wire and drives them in with great rapidity.

We have practically now a roughly formed shoe ready for the finishing room.

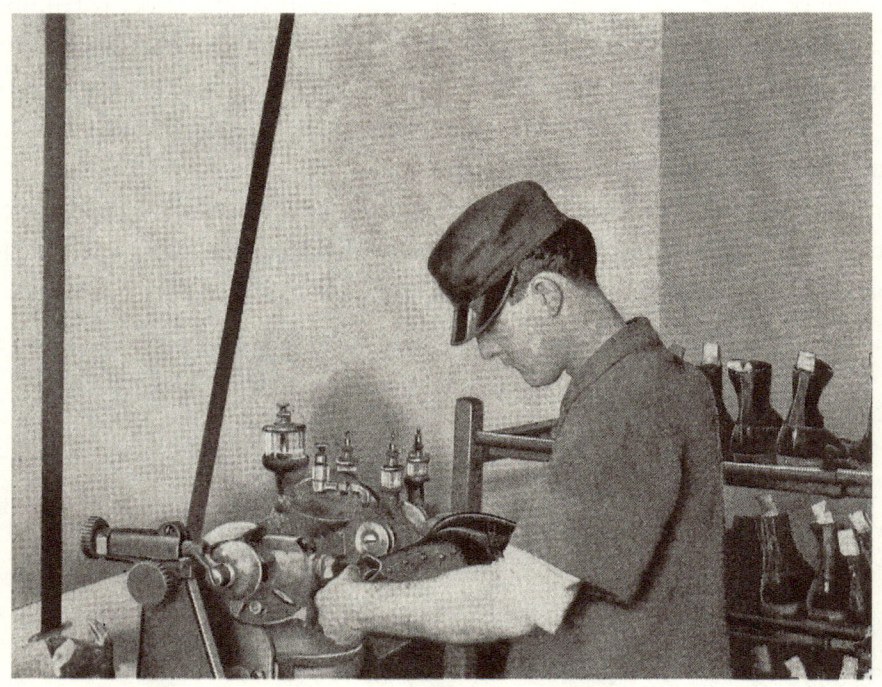

Heel Shaping.

Here the heel slugs are ground down, heel and sole buffed by sandpaper rolls on a scouring machine, wet down, stained, or blacked, as case may be, finished on bristle brushes, placed to dry, polished by a polishing machine, bottom stamped with the trademark, and passed to an operator whose duty it is to see that no tacks are left inside the shoes. Generally girls are hired to do this, as their hands are smaller and it is very important that no tacks are left, which might cause a great deal of trouble. If any are found, they are cut out with nippers or otherwise removed.

A lining is also generally put inside the shoe, covering the whole of the insole in a McKay shoe, and the heel only in a Goodyear shoe. Shoes must also be inspected here before they are packed, to see if they are perfect in every way and that each shoe is a perfect mate in the pair.

The shoes are now sent to the last department, called treeing, dressing, and packing department.

This department has to do with the finishing of the uppers. The bottoms and edges are all finished when shoes get to this department, and nothing remains but to finish the uppers and pack the shoes in single-pair cartons and then in wooden boxes or cases.

The different uppers are all finished by a different process, some being ironed with a hot iron, which is done to take out the wrinkles and smooth the uppers. Ironing was first introduced on kid shoes, but in recent years the hot iron has been put on nearly all kinds of stock. A shoe must be on a form or tree when ironed, the form or tree being the same shape as the last. The whole idea in ironing is the same as that followed by the tailor, who uses a hot iron to press and smooth out clothes. The operations in detail are as follows:—

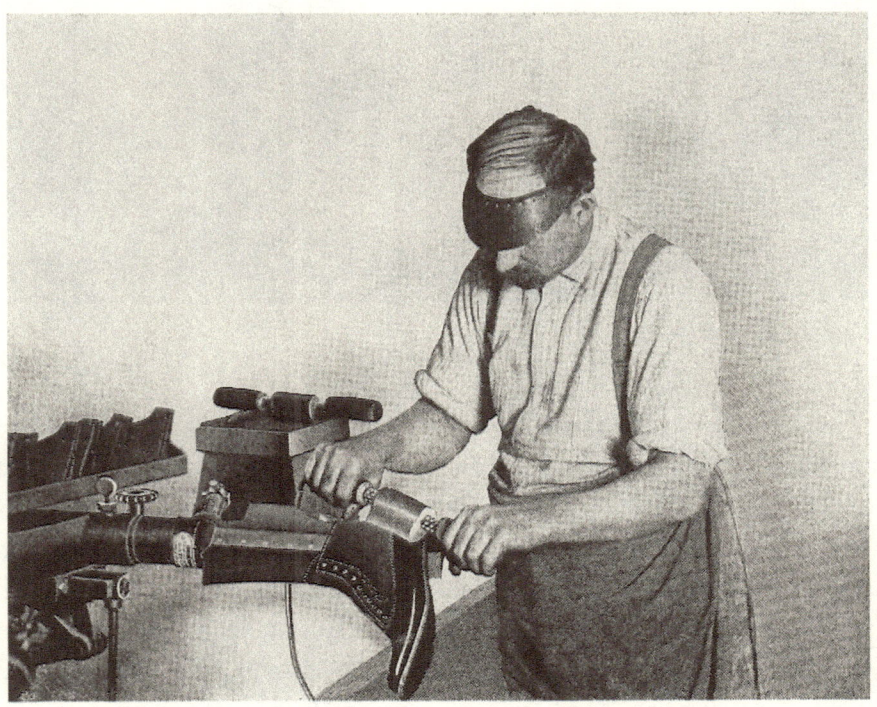

Ironing.

Each shoe is treed, after having been drawn over a foot form similar to that on which the shoe was lasted, and any stain or dirt which may have been carelessly put on in former operations is cleaned off; the shoe is sponged with a gum prepared for either black or tan goods, rubbed down dull, and then rubbed to a polish. In many patent leather shoes the treeing is to clean off the surface, as we said before, and then to iron it with a hot iron, which takes out all stains, and leaves the leather shiny and black.

Packing.

The shoes finally go to hand operators, who rag the edges and heels, leaving them ready to be laced and put into the boxes. After lacing, the shoes are passed to inspectors, whose duty it is to see that they are perfect, to throw out all which are not, make a record of them, and pass the perfect shoes to the packers, who see that the sizes are right, that each pair is mated, and placed in paper cartons, ready to be packed in wooden cases for shipment. The packing of cartons into wooden cases is done by men who nail on the lid when each case is full, mark where goods are to be sent, make a record of same and load the cases into freight cars.

There are other uppers that are treed, such as wax calf, for instance, and split uppers, which are used in heavy shoes. The main idea of treeing a shoe is to give it a smooth and finished appearance and a good "feel." In the regular treeing operation they use liquid preparations, often called composition, and these are worked into the upper, filling it to some extent. French chalk is used a great deal in some uppers, and oil or some form of grease or gum is also used, all of which make the upper as it was when first put on the cutting board of the shoe factory. All work done in this room is intended to give leather its original luster, which has been lost to a certain extent in going through the different rooms and in being handled so much.

There are still other uppers that may not be treed or ironed but merely cleaned and polished to give luster. Some of these may be dressed. To dress a shoe means to put on a liquid dressing. In some cases two coats of dressing are put on and in other cases one coat. A shoe can have a dull dressing or a bright dressing, according to how the prefers to have his shoes look.

# CHAPTER SEVEN.

## McKay and Turned Shoes

THE McKay process is used very extensively in the manufacture of cheap shoes. Its introduction was a great improvement over the nailing and pegging of the soles to the uppers. It allows the two to be stitched together by means of a straight needle running through the entire thickness of upper, sole, and insole.

In following the McKay process through the factory, we find it very similar to the Goodyear welt process, which has been explained, the main difference being in the methods of fastening the sole to the uppers.

The lasts and patterns are obtained in the same manner as described in the previous chapter. The order is made out in the factory office, and the ticket is given to the sorter, who selects the required number of skins, which he rolls in a bundle and turns over to the cutter. The cutters form the various pieces of leather and linings, which are tied up in bundles and sent to the stitching room. Here they pass through the various sewing machines, finally coming out in the form of a complete upper ready to be attached to the bottoms.

The soles, insoles, counters, and heels for McKay shoes are all formed in the same room, as described in the Goodyear process.

There is a difference in making ready the outsoles and insoles. It will be recalled that the outsole for the Goodyear welt shoe was simply a block of leather cut to fit the shoe and was not channeled. The outsole for the McKay shoe is run through a channeling machine, which cuts a slit around the edge of the sole, folds the leather back, and digs a little trench along the inside of the slit. It will also be remembered that the insole of the Goodyear welt shoe was channeled with two slits, one of which was turned back to form the breast for sewing on the welt strip. The insole of a McKay shoe is not channeled in any way, but is left plain, like the outsole of the Goodyear welt. The uppers, the soles, insoles, counters, and heels all having been made ready, the pieces are taken to the lasting room.

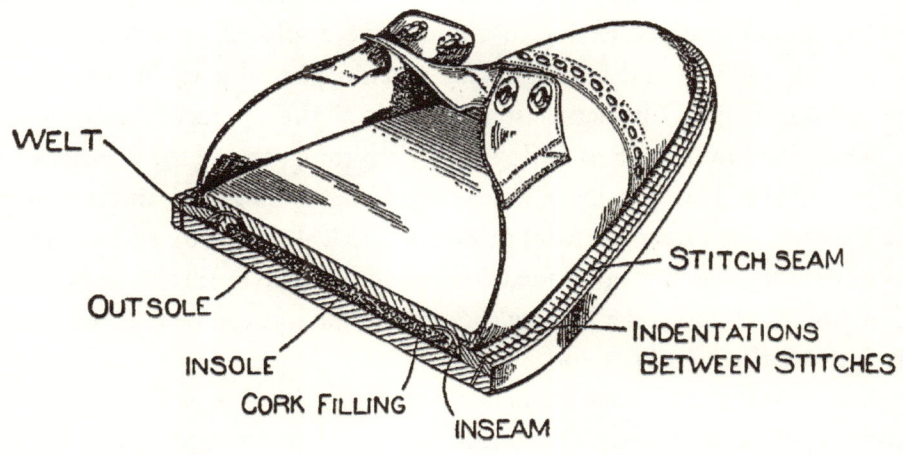

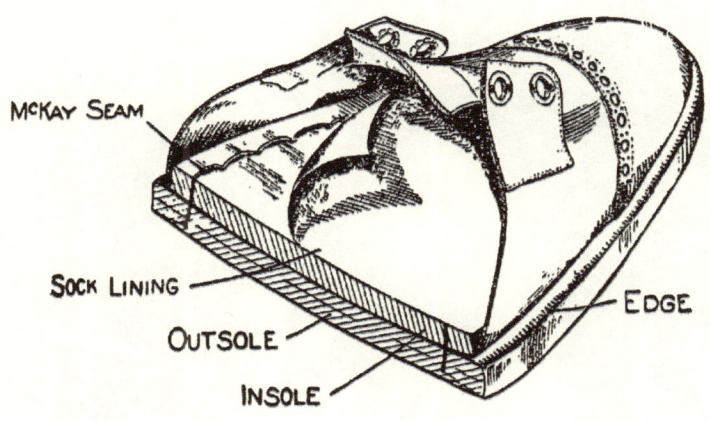

Cross Sections of Welt Shoe and McKay Sewed Shoe.

The first process is called "assembling." The operator takes up one of the uppers, inserts the last, sticks in a counter between the lining and the outside, puts in a "box" (a stout piece of canvas to give stability to the toe) at the toe, beneath the tip, puts in the insole, and then may pull the shoe tight on the last or give it to the operator on the pulling over machine to have it done. The pulling over machine is now used in nearly all factories, having displaced hand pulling the same as the lasting machines have displaced hand lasting. The assembling, pulling, and lasting on the machine are all parts of the regular operation of lasting. The hand laster had to do all three parts in former times, but now there are machines to do nearly everything, and at the present time the operation of lasting is divided into assembling, pulling over, and lasting on the machine. But even these machines do not do it all, as there is surplus upper to be cut away, toes to be pounded down, and filling to be put in the bottom, all of which are done on a McKay shoe before the sole can be laid. There are machines to do these parts, too.

A trimmer (this is done by hand) now takes the shoe, trims off all the surplus leather, tacks in the shank (a little piece of steel to give rigidity to the shank of the sole), fills all up smoothly and then passes it to the sole layer, who puts on the outer sole and tacks it in place.

Tacking.

The last is now pulled out of the shoe and it is ready for the McKay sewing machine.

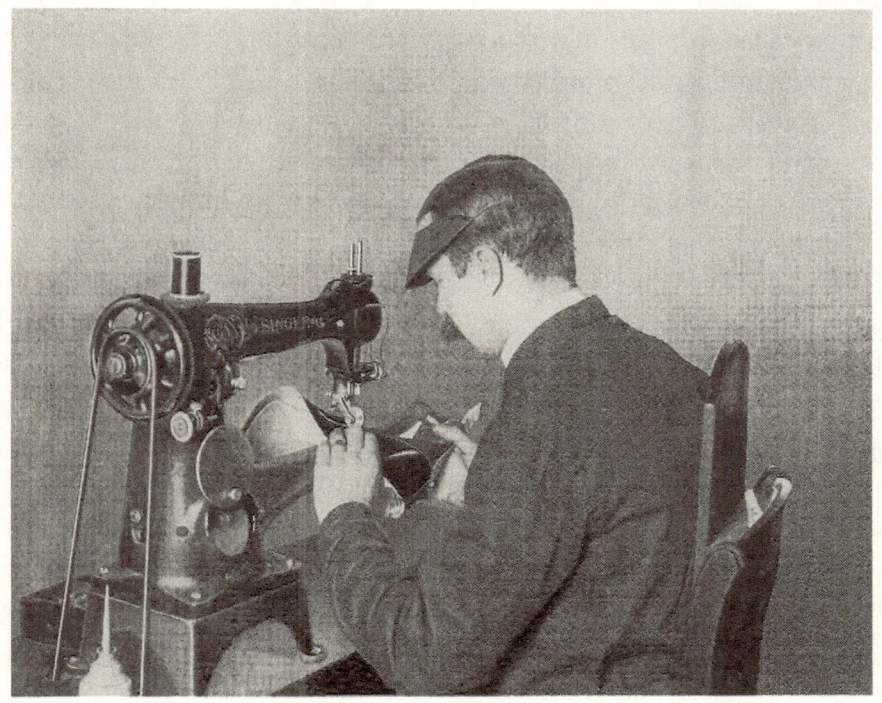

Stitching.

This machine sews right through the inner and outer sole, and at the same time catches the edges of the upper leather and the lining in between the two and draws them all snugly and firmly together. The stitches are made right along in the channel of the outer sole, which is deep enough to admit the row of stitches without raising a ridge on the outside of the sole, after the channel is closed up and leveled. The channel is next filled with cement and passed on to the leveler, which turns down the loosened flap of leather, presses it all out smooth, and covers the seam up so completely that no trace of the sewing is to be seen. This little folded-over flap of leather serves the double purpose of hiding the stitches in the sole, and at the same time protecting them from wear against the ground.

The shoe is then ready to be heeled, and from here to the shipping door the McKay generally goes through the same process as a welt. After heeling, the

McKay Shoes are relasted or have followers put in to keep them in shape while going through. The sock lining may be put in here, too, before relasting, or it may not be put in till the shoes get to another room. The McKay lasting last must be pulled from the shoe to have the bottoms and heels put on and this also applies to a pegged or nailed shoe. But in the case of a welt shoe or a turn shoe, both stay on the original last until the bottoms and heels have been fastened on. The turn shoe being lasted inside out, must come off the last to be turned right side out, and it goes right on the last as soon as it can be turned. The different methods of fastening the bottoms constitute the main difference between Goodyear and turn shoes on the one hand, and McKay, pegged, and nailed on the other. The bottom stock must be prepared differently in order to fit the methods. Thus it is seen that only two departments are affected, namely, the sole leather and the making departments. In the cutting, stitching, finishing, treeing, and packing, all operations are practically the same on every shoe, no matter how it is bottomed. The patterns, however, by which shoes are cut may be different.

In the finishing room all of the finishing of the bottoms and heel edges is done. The heels are sandpapered or scoured, and are then blacked and polished under hot-iron pressure. Considerable wax is used on the edge and is melted by the hot iron. Heel edges may also be finished on a wheel or roll. There are several different ways, but the object of each method is to give a hard, black, and highly polished surface to the edge.

In finishing the bottom the top lift is scoured or buffed, and all of the sole and the breast of the heel also. Each is a different process, a different operator attending to each part. The object of scouring or buffing with sandpaper is to get a smooth foundation for the finish, which is put on next, and which may be all the same color in all parts of the bottom or may have one color in the shank and another in the fore-part. The stains and blackings are used on bottoms, and these are brought to a high, hard gloss by means of rolls and brushes. Hot irons are often used on black shanks and bottoms to give added hardness and luster to the finish.

The turned or turn shoe is a woman's fine shoe that is made wrong side out, then turned right side out. The sole is fastened to the last, and the upper is twisted over, the wrong side out. Then the two are sewed together, the thread catching through a channel or shoulder cut in the edge of the sole. The seam does not come through to the bottom of the sole, nor to any part inside where it would chafe the foot.

The preparation of the upper for a turn shoe is identical with that of a welt or McKay, with the exception that the back is cut a little longer and a little larger, in order to last it over the sole. The important difference in the make-up of a turn shoe as compared with that of a McKay or welt is that it has no insole, the upper being sewed directly to a portion of the sole itself.

As the cutting of the uppers and the stitching operations of a turn shoe are the same as the Goodyear and McKay, and have been explained, we will take up the forming of the sole, which is entirely different from either of the other two methods.

A turn shoe is put together wrong side out, and it is necessary, during the course of making, to turn it by rolling the sole up like a roll of carpet. It is evident, then, that nothing but good quality, pliable leather can be used satisfactorily, and great care is taken to include nothing but the best.

The soles are cut out on the beam machines, also previously described. They are then channeled on the side that is next to the foot. This channeling is similar to that done on the welt insole. Two incisions are made, the inside one being the same as in the welt insoles. The outside one, however, is different, as the flange is cut off square instead of being rolled up. This leaves a channel which begins at the edge and surface of the sole and extends in semicircular form to the abrupt wall of the cut in the sole, which forms the breast against which the upper is to be sewed.

After the soles are channeled, they are soaked until they become soft enough to roll up easily. They are then placed on racks and kept in a damp room until needed.

A turn shoe is hand lasted wrong side out. First the uppers are turned with the lining outside, then the last is inserted and also the toe boxing.

The sole is set straight on the last and is tacked firmly to it. The operator, by aid of hand pullers, draws the upper over the sole and tacks it securely from a point where the breast of the heel will rest to where the large toe will extend, and then along the same distance on the other side. The toe part is next lasted by machinery, a wire being fastened at one side and run around the edge holding the pulled-up parts of the upper which has been stretched tightly over the last.

The shoe is next passed over to the Goodyear in-seamer operator, who sews the upper to the sole, the needle passing down through the inside channel, through the sole leather, out through the square-cut channel and then through the upper, uniting the upper to the sole with the chain stitch. In fact, the bottom of a turn shoe at this time looks exactly like the bottom of a welt, with the exception that the turn shoe is still turned wrong side out. The nature of the stitch is the same—a waxed, threaded chain, with two rows of thread on the outside that loop with the single thread in the inside lip of the insole. The shoe is sewed only from the back of the shank to the toe, the heel part still being loose.

The seam is now trimmed with an inseam trimmer, a machine with a revolving, jagged-edged knife that saws off the surplus portions of the upper, leaving it smooth and even with the sole. The tacks are all pulled out with a sort of a nail puller, which works rapidly and automatically.

The lasts are then taken out and the shoe is turned right side out. This turning process is not a difficult one, but it is perhaps the most interesting operation that the layman will see in the entire factory. The operation is accomplished by means of a rigid iron bar set slantwise in a table. The upper is turned right side out by hand and the sole is rolled right side out by means of pressure on this bar.

After this turning process, which twists and rolls the shoe out of shape, it has no semblance of its final form. The back part of the sole and upper are still loose, the upper being fastened from the shank to the toe.

The turn shoe must be "second" lasted, and the inserting of the last is no easy matter. A contrivance called a push jack assists the operator greatly. He uses a flat, narrow rod to smooth out the lining, and after squeezing, pushing, and smoothing, the last is finally made to fit in the shoe. The counter is placed in at this time, the shank piece is set in place, and the shoe and last are placed on a jack for nailing. The back part upper is now stretched tightly over the heel part of the last by means of lasting pullers, and is tacked down, the nails going through the shank piece and clinching against the anvil heel seat of the last. This operation completes the lasting, the shoe now having a form exactly like the last over which it is made.

Workmen now level the bottoms and form the shank by a hand method, preparatory to the machine leveling process. The shoe is still wet and is left to dry on the last twenty-four hours. Then it is run through the machine called the "leveler," which, with its enormous pressure, forms the sole to that of the last. The shoes are now left four days on the lasts, to dry thoroughly, so that they may retain their shape permanently.

The putting on of the heel, and the various finishing processes are practically the same as that of the welt, with the exception that a turn sole must have a sock lining.

Some factories use a grain leather sock lining, which is pasted in, covering up the channels of the sole which hold the stitches and forming a smooth surface for the foot to rest upon.

The difference between a McKay and a turn shoe may be told by the fact that the stitching on the inside of the sole is much closer to the edge in a turn. Another thing, in a turn shoe, the seam connecting the upper and the outsole can be seen.

Nothing is likely to excel the turn shoe for lightness and flexibility, since the method of making, whereby the sole is stitched directly to the upper, interposes no thick or cumbersome material. Sole leather of good quality is used. In fact, the sole would have to be not only strong, but thin and light, or the shoe could not be turned in the process of manufacture without straining it and getting it out of shape.

### History of the Turn Shoe

History states that prior to 1845, which marked the date of the introduction of shoe machinery, most of the shoes were sewed by hand, the lighter ones turned and the heavier ones welted. In fact, the early factories that began to spring up in New England about the beginning of the century, were merely cutting rooms and places for storing the lasts and stock.

Here the uppers, soles, and linings were cut by hand and then given out to people in the vicinity, mostly farmers, and fishermen, to be stitched together and paid for at so much a dozen. Such was the beginning of the shoe industry in New England. Hundreds of families added to their resources in this way, the women doing the lighter work and the men the heavier.

In fishing communities, where men were away most of the time in their boats, their wives and daughters, who stayed at home, undertook the lighter grades of shoemaking—the turn process. This was the case in the "North Shore" towns like Lynn, Haverhill, and Marblehead, and these to-day, keeping to the old traditions, are the great centers for the finer turn-grades of shoemaking, whereas the "South Shore" towns, like Brockton, Whitman, Abington, Rockland, and the Weymouths, with the men at home all the year, came to make a specialty of shoes for men, and absorbed the heavier part of the growing industry.

With the introduction of the Goodyear turn machine, however, the handwork was gradually done away with, although more handwork is done in the turn process than in either the McKay or welt process.

### STANDARD SCREW SHOEMAKING

Many good qualities of heavy shoes are made by the standard screw method, which differs from the McKay method by having the outsole and insole fastened together with a double-threaded wire, which is screwed through and cut off by the machine the instant it reaches the inside of the shoe.

A pegged shoe is made in much the same way as the standard screw, except that wooden pegs are used instead of wire to fasten the sole together.

The nailed method of shoemaking consists in nailing the soles together around the edge. It is used principally for heavy, cheap shoes.

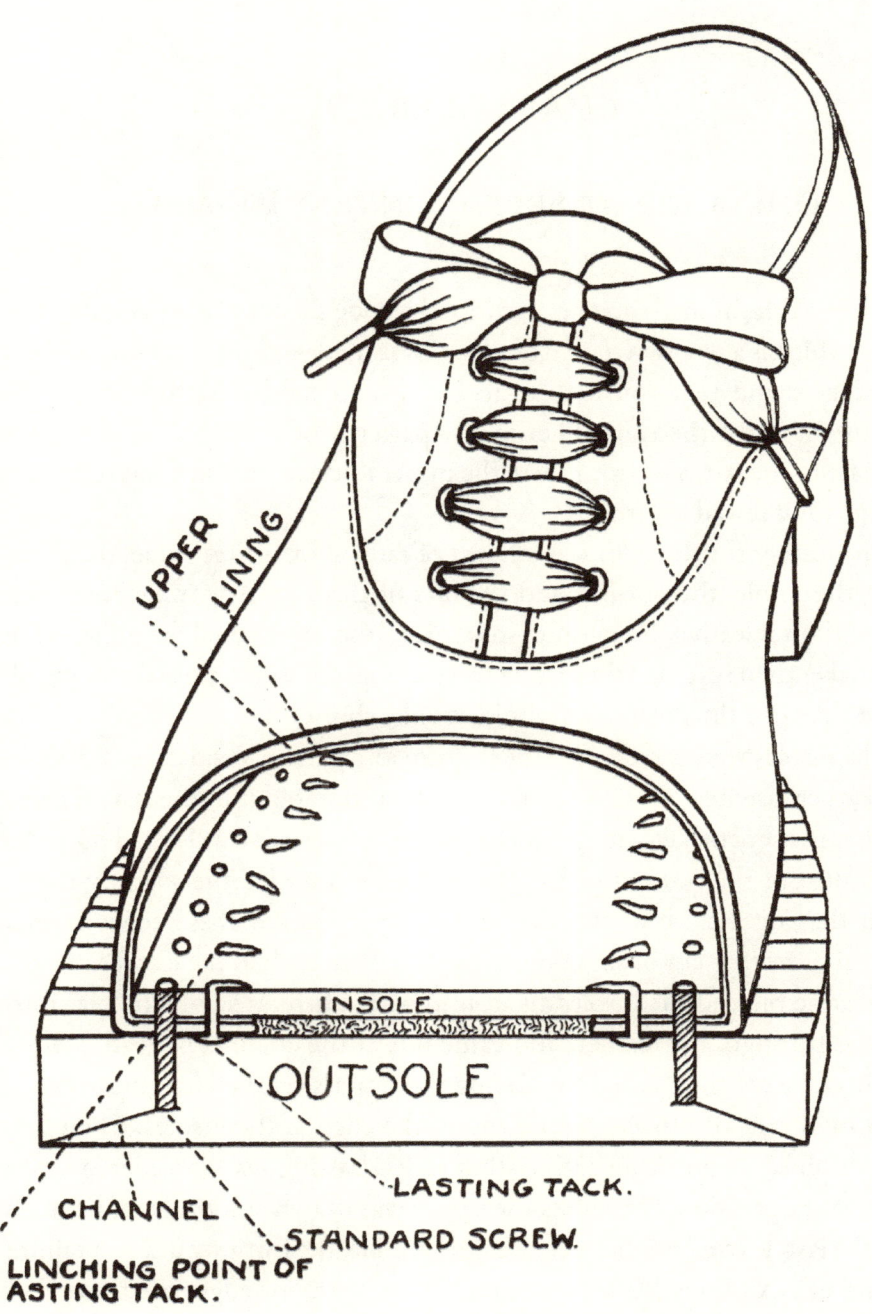

Cross Section of Standards Screwed Shoe.

# CHAPTER EIGHT.

## OLD-FASHIONED SHOEMAKING AND REPAIRING

THE old-fashioned shoemaker formerly made shoes by hand as follows:—A last, which is a wooden model of a foot, was used, and pieces of leather were pasted here and there on it so as to build up a model conforming to the measurements of the foot. Then paper patterns of the upper leather were made from the last, and from these the upper leathers were cut out of tanned calfskins and sewed together.

The leather for the soles was cut out of tanned ox or steer hide, the pieces being the insole, the outsole, and the lifts of the heel. The inner soles were made of softer leather. Sometimes split sole leathers were used for uppers. The shoemaker then softened the leather by steeping it in water, until it was pliable and at the same time firm, and would cut like cheese.

The insoles were attached to the bottom of a pair of wooden lasts, and the wet leather fastened on with lasting tacks so as to mold it to the last. When it was dry, the shoemaker with pincers drew the leather out until it had taken the exact form of the bottom of the last. Then he rounded the soles by paring down the edges close to the last, and formed around these edges a small channel or feather cut or slit about an eighth of an inch in the leather.

Next he pierced the insoles all around with a bent awl, which "bit" into, but not through, the leather, and came out at the channel or feather edge. The boots were then lasted by placing the uppers on the lasts, drawing the edges by means of pincers tightly round the edge of the insoles. Then they were fastened in portions with lasting tacks. Lasting was considered a very important operation, for unless the upper was drawn smoothly and equally over the last, leaving neither a crease nor wrinkle, the form would be a failure. A band of flexible leather about an inch wide, with one edge pared, was then placed in position around the sides of the shoes, up to the heel or seat, and the maker proceeded to "inseam," by passing his awl through the holes, already made in the insole, catching with it the edge of the upper and the thin edge of the welt, and sewing all three together in one flat seam, with a waxed thread.

The threads which shoemakers use are called "ends," and are made of two or more strands of small flaxen threads. The shoemaker makes his own waxed thread as follows:—

He holds the main part of the thread from the spool, in his left hand, holding it firmly—where he wants to break it—between the first finger and thumb, so that it will not turn beyond that point. Then with the left hand, he lays the end of the flax on the knee and rolls it from him. This will cause the small fibers that compose the thread to separate—thus enabling him to break it easily. When the fibers separate, he gives the thread a light, quick turn, which causes it to break. As the thread breaks he pulls it apart gradually, so that the fibers will taper. Then he places the threads together, one just behind the other, so that the end will have a very fine point. He rolls the end and allows it to turn between the fingers of the left hand. After it has been rolled and twisted, it is waxed by drawing the thread through a piece of wax.

The fine ends are waxed to a point. A bristle is fastened on in the following manner: the head of the bristle is held in the left hand, and the portion to which the thread is to be fastened is waxed; then the thread and bristle are twisted together. A hole is made in the thread and the bristle pulled through and fastened. After the threads are fastened, the heads of the bristle are cut off, and the ends sandpapered.

The wax thread or "end," as it is called, should never be made longer than is necessary to sew a shoe. Experience shows that if a portion of an end left after sewing one shoe is used on the second shoe, it is never as strong as a new end. The thread grows weaker and weaker as it is used. When the thread is well waxed, it is cemented to the shoe.

After the shoe is sewed, the shoemaker pares off inequalities and levels the bottoms, by filling up the depressed part in the center with pieces of tarred felt. The shoes are now ready for the outsoles. The fibers of the leather to be used for the soles are thoroughly condensed by hammering on the lapstone. Then they are fastened through the insole with steel tacks, their sides are pared, and a narrow channel is cut round their edges. Through this channel they are stitched to the welt, about twelve stitches of strong, waxed thread being made to the inch. The soles are next hammered into shape; the heel lifts are put on and attached with wooden pegs. Then they are sewed through the stitches of the insoles; and the top pieces, similar to the outsoles, are put on and nailed down to the lifts.

The finishing operations of the shoe include smoothing the edges of the heel, paring, rasping, scraping, smoothing, blacking, and burnishing the edges of the soles, withdrawing the lasts, and cleaning out any pegs which may have pierced through the inner sole. There are numerous minor operations connected with forwarding and finishing in various materials, such as punching holes, inserting eyelets, etc.

### How Shoes are Repaired

Before one can understand how shoes are repaired, it is necessary to know the difference between the inside and outside of a shoe.

The last is divided into four parts, viz. toe, ball, shank, and heel.

Diagram No. 1 shows these parts and their shapes.

Diagram No. 2 shows the length of the inside of the divisions as compared with those of the outside. Notice the long shank and short ball.

Diagram No. 3 shows the outside of the divisions and the effect they have upon the shape of the shoe. See short shank and long ball.

Always remember that the ball of a shoe is longer on the outside, having a short shank. The ball is shorter on the inside, having a long shank. Compare outside and inside diagrams Nos. 2 and 3.

### Shoe Repairing

The first operation in half soling a shoe is to cut off the old portion from "a" to "c" as shown on diagram No. 1. The shoe is placed in different positions and corrected in every way before putting on the new sole. It is generally better to wet the shoe in order to put it in shape.

The leather is skived thin and accurate enough to make a neat, comfortable joint, and yet thick enough for the nails to hold.

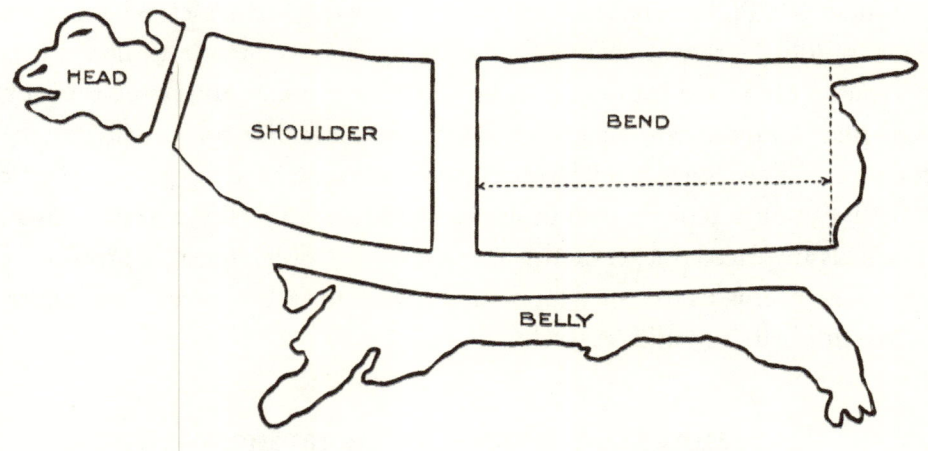

How a Side of Leather is shaped and divided as to Quality. *See Page 5.*

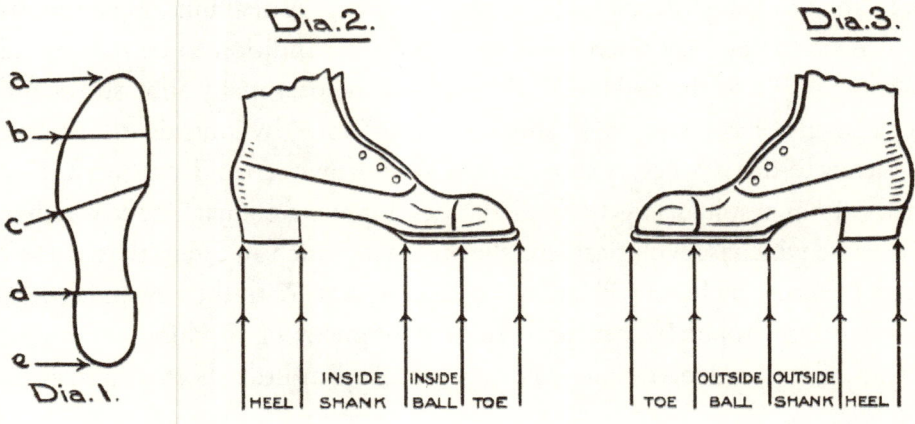

Then the filling is added before placing on the sole. The sole is trimmed and a guide line drawn around the edge, so that the nails may be properly arranged.

Finishing the sole is an important part. If everything else is properly done, this part becomes comparatively easy. See that all nails are clinched. With a level bottom, smooth joints and edges, the shoe can be made to look like a new shoe and yet feel like an old one.

On account of the heel being more directly under the body and the first part to strike the ground, it generally wears out first. For this reason in repairing a heel great care must be taken to see that good leather and solid work are put into it. Pull off the worn top piece and see that what is left is hammered down solidly. Next split a piece of solid, easy-cutting, scrap sole leather, so that two pieces can be made out of one. Put them on the shoe and fasten them on well, piece by piece, with tacks. See that the heel is level before putting on the top piece. (If necessary, a small piece may be put under the top piece.) After it is level, put on top piece, trim in shape, then draw guide line and nail down. The nails are placed thicker on the side that is worn down most, to protect the heel. The heel is next rasped, and smoothed with a buffer and sandpaper. When finished, it should set level.

## Modern Method of Repairing Shoes

As the shoemaking industry has become more and more perfect, there has been an increasing interest taken in shoe repairing. A medium-priced shoe as it is made to-day may often be in good enough condition to be heeled and soled a couple of times. Hence, although in the past many shoe stores and departments have had their shoe repairing done by outside shops, the tendency to-day is for every shoe store to have its own repair department. This method has resulted largely from the development of machinery for shoe repairing, which is revolutionizing the business to such an extent that in a few years repairing by hand will be among the lost arts. With the new inventions for restoring upper leather, and the improvement of machinery for shoe repairing, repair departments will very soon be but little short of miniature factories.

The machinery ordinarily used consists of the Goodyear stitcher, used for attaching soles to Goodyear welts by the lock-stitch method, just as in shoe factories making Goodyear welt shoes. Then there is a heel trimmer, a bottom finisher, consisting of a rapidly revolving roll covered with coarse and fine sandpaper, and an opera heel builder for forming concave heels. There are two wheels used for tan and white heel work, one heel being covered with a white cloth, and the other with a coarse brush. Adjoining these are usually the shank and heel finisher,—capable of smoothing and highly polishing a shank or heel in about a dozen seconds,—the bottom finisher, that grinds and smooths down the new sole, and a machine used for rubbing off dirt before the shoe is finished, consisting of a heavy horsehair brush. Another useful part of the equipment is an edge setter, which is also identical with the one used in factories. The shoe stitching machines and the parts used in finishing are all operated on one long shaft, rapidly revolved by the aid of a motor. It is a fact that a shoe may be actually soled and heeled in less than six minutes.

Five or six men are usually employed in the repair department of a large establishment. When the customer's shoes are brought in, one of these men cuts off the old sole and traces an outline of the new sole on a block of the very best oak leather. After these are cut out by hand in rough form, they are soaked in water and channeled that is to say, a part of the sole is turned up in which the stitches are to be run. A second man, by the use of the Goodyear stitcher, joins the sole and welt together with a very strong and tightly drawn lock stitch. This is a large machine with a curved, barbed needle and awl, and a shuttle which sews through an inch of leather with the greatest ease and speed. There are from one hundred and fifty to two hundred stitches in each shoe; moreover, every one of them is locked with heavy wax thread, so that there is no chance of their ever giving away. If one stitch should break, the other stitches would remain intact, as they are all independent of each other. Both soles are stitched on in a little over half a minute without breaking a thread or stopping the machine.

A coating of rubber cement is now placed in the edges of the outsole, and the lip of the channel is smoothed down so that the stitches are entirely hidden when looking at the bottom of the shoe. The edge trimming is done next with the aid of a rapidly revolving wheel, which trims the edges square and true in about forty seconds. After this, the shank is finished on a rapidly revolving wheel covered with emery cloth.

Bottom finishing is the next step. This is done on a machine having two long cylinders, one covered with fine and the other with coarse sandpaper. These cylinders revolve rapidly, and the operator uses the coarse sandpaper for scouring the dirt and old finish off the leather, and the fine sandpaper for finishing the sole as smooth as that of any new shoe.

The brushing in or smoothing is next done by the horsehair brush we have mentioned before. A preparation called Lewis's rival bottom polish—a sort of white wax—is placed on the brush machine. The brush now smooths the surface of the sole, filling in all small holes with wax and leaving the sole absolutely perfect. Finally, the shoe is placed against a rapidly revolving brush which finishes the uppers with a luster that would make any ordinary bootblack green with envy. Another operation that fully completes the process is the hardening of the edges with hot steel, which ends in producing an edge that is as hard as iron. When it is polished with a black dye, it looks exactly like a new sole.

A few words are necessary with regard to the heel. The old heel having been removed, several lifts of new leather in rough form are tacked on. The shoe is then taken to the heel trimmer and is formed correctly and then smoothed down to a brilliant surface on the finely covered revolving wheel. In a few seconds it is stained, smoothed, and polished. In less than six minutes the shoe is ready for the customer.

# CHAPTER NINE.

## LEATHER AND SHOEMAKING TERMS

ASSEMBLING. Includes the following operations: tacking the insole to the last, putting in the box and counter of the shoe, and putting the upper of the shoe on the last.

BACKSTAY. A term used to denote a strip of leather covering and strengthening the back seam of a shoe. English backstay means the strip of leather that meets the quarters on each side and is sewed to them, forming the lower part of the shoe. California backstay is a term applied to piping caught in the back seam.

BACK STRAP. The strap by which the shoe is pulled on the foot.

BAL. An abbreviation of the word "Balmoral" and means either men's, women's, or children's front lace shoe of medium height, as distinguished from one that is adjusted to the ankle by buttons, buckles, rubber goring, etc.

BALL. Refers to the ball of the foot—the fleshy part of the bottom of the foot, back of the toes.

BEADING. Means folding in the edges of the upper leather instead of leaving them raw, or wheeling any impression around the sole to the heel. It is called seat wheeling in many shoe factory rooms.

BEATING OUT. The same as leveling. It is the term used in turn-shoe work.

BELLOWS TONGUE. A broad tongue sewed to the sides of the top, seen in waterproof and some working shoes.

BELTING. The term applied to the usual back tanned cowhide, used in various thicknesses for machinery belts.

BETWEEN SUBSTANCE. That part of the sole that holds the stitch.

BLACKBALL. A mass of grease and lampblack, formerly used by shoemakers on edges of heels and soles; sometimes called "cobbler's botch."

BLACKING THE EDGE. Blacking or dyeing edge of sole, welt, or that part of the edge which cannot be blacked so well in the making room.

BLOCKING. The cutting or chopping of a sole in such a form or shape that it can be rounded.

BLOOM. A term often applied to the grayish white deposit that gathers on shoes in stock. It can be wiped off readily.

BLUCHER. The name of a shoe or half boot, originated by Field Marshal Blücher of the Prussian Army, in the time of Napoleon I. It became very popular and has since received occasional favor, being used with high tops as a sporting or hunting boot. Its distinguishing feature is the extension forward of the quarters to lace across the tongue, which may be an extension upward of the vamp.

BOOT. A term used (especially abroad) to designate women's high-cut shoes. In this country it applies only to high or topped footwear, usually made with the tops stiff and solid. It is sometimes laced, as in hunting boots.

BOOTEE. Leather legging extending between knee and ankle, usually of Russian calf,—a riding boot originating with the English.

BOTTOM FILLING. The filling that goes in the low space on the bottom in the forepart of the shoe. It is either ground cork, tarred felt, or other filler.

BOTTOM SCOURING. Sandpapering the parts of the sole, except the heel.

BOXING. A term used to designate the stiffening material placed in the toe of a shoe to support it and retain the shape; such as leather, composition of leather and paper, wire net, drilling (a cotton fabric) stiffened with shellac, etc.

BOX CALF. A well-known proprietary leather having a grain of rectangularly crossed lines.

BOX TOE. Used to hold up the toe, of the shoe so as to retain the shape. It is generally of sole leather, but often made of canvas or other material and stiffened with shellac or gum.

BREAKING THE SOLE. Molding the sole so as to fit the spring better.

BROGAN. A heavy pegged or nailed work shoe, medium cut in height.

BRUSHING. The final finish of the top edge, heel, and bottom, by means of a brush.

BUCKSKIN. A soft leather, generally yellow or grayish in color. One way of preparing it is by treating deerskins in oil.

BUFF. A split side leather, coarser than glove grain, but otherwise similar. It is used for cheaper grades of shoes, principally for men.

BUFFING. The same as bottom scouring.

CABARETTA. A tanned sheepskin of superior finish used for shoe stock. There are sheep with wool not far removed from hair in texture, which produce a skin of greater tenacity and finish than the ordinary sheep.

CACK. A sole leather bottom without a heel. An infant's shoe is called a cack.

CALFSKINS. Skins of meat cattle of all kinds, weighing up to fifteen pounds, are usually included in this term. They make a strong and pliable leather. Calfskins were formerly finished with wax and oil on the flesh side, but can now be made so as to be finished on the "grain," which is the hair side of the skin.

CAP. A term meaning the same as tip.

CARTON. A cardboard box intended for one pair of shoes.

CEMENTING. This is the operation of placing cement on the outsole and the bottom of the welt shoe so that the outsole is held to the shoe by the cement.

CHAMOIS. A leather made from the skins of Chamois, calves, deer, goats, sheep, and split hides of other animals.

CHANNELING. Cutting into the sole in such a way that the thread or stitching is away from the surface. In the outsole department it means preparing a place for the stitch. In insoles and turn soles, channeling is done so that soles are prepared to hold the stitching.

CHANNEL SCREWED. A process by which the sole is fastened to the uppers. After a channel is cut and laid over on the outside of the outsole, the outsole and insole are fastened together, holding the upper and lining between them by means of wire screws, which are fastened in this channel. The skived part is then smoothed down over the heads of the screws, entirely covering them from sight, and preventing the screws from easily working up into the foot.

CHANNEL STITCHED. A method of fastening soles to the uppers, either by McKay or welt process, in which a portion of the sole's outer side is channeled into, and the stitches afterwards covered on the lower side by the lip of this channel.

CHANNEL TURNING. Turning a lip or flap of sole leather (called channel), so that the stitching can be done in the proper place; or it may mean turning up the flap or lip of the channel, that is, the part that is to cover the stitch.

CHECKING. A term applied to the edges of heels or soles that have cracked, or have been injured in process of construction.

CLEANING INSIDE. Cleaning the lining.

CLEANING NAILS. Scraping the blacking off the tops of the heel slugs.

CLEANING SHOES. Removing dirt, wax, cement, etc., from them.

CLICKING. Cutting the uppers of shoes.

CLOSING. Putting two or more pieces together.

CLOSING ON. Stitching the lining and outside together.

COLONIAL. A name given to a woman's low shoe, with vamp extended into a flaring tongue, with a large, ornamental buckle across the instep. The buckle and tongue are the distinctive features of the shoe, whether the shoe fastens with a lace or strap.

COLTSKIN. Coltskin has been brought into general use in shoemaking within the past few years. The skin of a colt is thin enough to use like calfskin in its entirety, with such shaving as is given all hides in tanning. Coltskin makes a firm basis needed for patent leather, and has been much used in recent years for this purpose. Russia is the chief source of supply.

COMBINATION LAST. One with a different width instep from the ball. It may be one or two widths' difference, such as the D ball with a B instep. Combination lasts are generally used in fitting low insteps.

COMPOSITION. A term used to denote the small scraps that accumulate about tanneries and factories, which are ground up and mixed with a paste or a kind of cement, and flattened into sheets which are used as insoles, and in other parts, in various grades of shoes, where wear is not excessive.

CONGRESS GAITER. A shoe designed especially for comfort, with rubber goring in the sides which adjusts it to the ankle, instead of laces, and sometimes made with lace front to imitate a regular shoe.

CORDOVAN. Originally a Spanish leather made from horsehide. The Spaniards were, for a great many centuries, the best leather makers. The term is applied to a gram leather from the best and strongest part of a horsehide.

COUNTER. The stiffening in the back part of a shoe, often called stiffening, to support the outer leather and prevent the shoe from "running over" at the heel. It is made either of sole leather, shaved thin on the edge and shaped by machinery, as in the best shoes, or composition or paper, in cheap shoes. Metal is occasionally used on the outside of the shoes in heavy goods for miners and furnacemen.

COUPON TAG. A tag from which a coupon is cut for every operation. Operatives hold part of the coupon and the holders of the coupons are paid for the part named.

COWHIDE. Refers to hides of cattle, heavier than kips, which run up to twenty-five pounds each.

CREASING VAMP. Making hollow grooves across the front of the vamp to add to its looks.

CREEDMORE. A man's heavy lace shoe, with gusset, blucher cut.

CREOLE. A heavy congress work shoe. This shoe, the creedmore, and brogans are usually made of oil grains, kip, or split leather, sometimes pegged, sometimes "stitched down."

CRIMPING. Shaping any part of the upper so that it will conform to the last better.

CUSHION SOLE. An elastic inner sole.

CUT-OFF VAMP. One cut off at tip for economy when tip is to be covered by a cap.

DIEING. Cutting soles to fit the last, outsoles, insoles, heel lifts, counters, or half soles, with a machine and a die.

DOM PEDRO. A heavy, one-buckle shoe, with gusset or bellows tongue. Originally it was a patent name for certain shoes made of fine material, but is now applied to cheap grades.

DONGOLA. A heavy, plump goatskin, tanned with a semibright finish.

DRESSING. A process for giving the upper its original finish by means of liquid put on with sponge.

EDGE SETTING. The finishing edge of the sole,—polishing it.

EDGE TRIMMING. Trimming the edge of a sole smoothly to conform to last.

ENAMEL. Leather that is given a shiny finish on the grain side. The process is similar to that of patent leather, only that patent leather is finished on the flesh side, or the surface of the split.

EYELET. A small ring of metal, etc., placed in the holes for lacing; the eyelet holes are sometimes worked with thread like a buttonhole.

EYELETTING. Putting on eyelets.

FACING. The bleached calf or sheep skin used around the top of the shoe, and down the eyelet row and inside of the upper.

FAIR STITCH. Term applied to the stitching that shows around the outer edge of the sole, to give the McKay shoe the appearance of a welt shoe.

FAKING. Putting a gloss on any part of the bottom of the shoe.

FINDINGS. The small parts of a shoe, such as blacking, cement, nails, wax, tacks, thread, etc.

FLAP, LIP, AND SHOULDER. Terms used in connection with the channel or with the operation of sewing.

FOLLOWER. Any last or form put in a shoe from which the original last has been pulled.

FOREPART FINISHING. The staining and polishing of the forepart of the shoe.

FORM. A term applied to a filler last. It may be of wood, papier-mâché, leather board, or any similar material, and is used to enhance the appearance of sample shoes, in salesmen's lines or in window displays.

FOXED. Having the lower part of the quarter a separate piece of leather or covered by an extra piece; "slipper foxed" is a term sometimes applied to women's full vamp shoes.

FOXING. The name applied to that part of the upper that extends from the sole to the laces in front, and to about the height of the counter in the back; being the length of the upper. It may be in one or more pieces and is often cut down to the shank in circular form.

FRIZZING. A process to which chamois and wash leather are subjected, after the skins are unhaired, scraped, "fleshed," and raised. It consists in rubbing the skins with pumice stone or a blunt knife till the appearance of the grain is entirely removed.

FRONT. A term used for part of a congress toe.

GAITER. A term usually applied to a separate ankle covering or to a congress shoe.

GEMMING. The Operation of making gem insoles.

GEM INSOLES. An insole for welt shoes of leather.

GLAZED KID. See Kid.

GLOVE GRAIN. A light, soft-finished, split leather, for women's or children's shoes or topping.

GOATSKIN. See Kid.

GOODYEAR WELT. A term used to denote the process of attaching the sole to the upper of a shoe by means of a narrow strip of leather called a welt.

GORE. A rubber elastic used in a congress shoe. It is also applied to the long, wedge-shaped piece of leather set in an upper to widen it.

GRADING. The sorting of outsoles and half soles to get uniform weight in edges of finished shoes.

HALF SOLE. Half of a complete sole used in forepart of bottom under outsole.

HARNESS LEATHER. Similar to belting, and is made from hides heavier than kips.

HEEL. Made of layers of leather or wood called liftings, and attached to rear part of shoe (heel seat). There are different varieties of heels. The French heel is an extremely high heel with a curved outline in back and front (breast). It is sometimes made of wood covered with leather, with thicknesses of sole leather, or all sole leather. The Cuban heel is a high, straight heel, without the curve of the French or "Louis XV" heel. Military heel is a straight heel not as high as the Cuban. A spring heel is a low heel formed by extending back the outside of the shoe to the heel, with a slip inserted between the outsole and heel slat. Wedge heel is somewhat similar to a spring heel, except that a wedge-shaped lift is tacked on the outside instead of a slit. Slugging heels is the process of affixing the made-up heel by one operation of the machine.

HEEL FINISHING. Blacking and polishing the heel edge.

HEEL LINING. The lining to cover heel nails inside the shoe; it is often known by other names.

HEEL PAD. In the manufacture of shoes, is a small piece of felt, leather, or other substance fastened to and covering the full width of the insole at the point upon which the heel rests. A heel cushion is sometimes called a heel pad.

HEEL SCOURING. Sandpapering the edge of the heel, except the front or breast portion.

HEEL SEAT. That part of sole on which heel is fastened.

HEEL SEAT NAILING. Nailing the heel part of sole.

HEEL SEAT TRIMMING. Trimming the rear or heel part of sole.

HEEL SHAVING. Shaving the heel, shaping it.

HEMLOCK TANNED. A process of tanning leather by hemlock bark.

HIDES. Distinguished from skins, in the trade. Hides refer to skins of animals which are over twenty-five pounds in weight. Skins refer to smaller animals; as skins of goats, calves, sheep.

INLAY. A trimming of the upper by an insertion of the same or different kind of material than that of the body in which it is inlaid. It is used for decorative purpose on a shoe.

INSEAMING. Sewing sole on turn shoe. Welting and inseaming are practically the same operation.

INSEAM TRIMMING. Cutting off the surplus leather; term is also applied to pulling sole tacks.

INSOLE. The first sole laid on the last, and is the foundation of all shoes with insoles. It is an important though invisible portion of a shoe. This inner sole is the part to which the upper and outsole are sewed or nailed in the McKay and welt shoes.

INSPECTING. The examination of shoes to see that the work is perfect; it is sometimes called crowning.

INSPECTING INSOLE. The operation of looking inside of the shoe for tacks.

INSTEP. The top of the arch of the foot.

IRON. A term indicating the thickness of sole leather; each unit is approximately one thirty-second of an inch in thickness.

IRONING UPPERS. Taking wrinkles out of the uppers and smoothing the same with a hot iron.

JULIETTE. A woman's house slipper which is cut a little above the ankle in front and back, and cut down on the sides is called a Juliette.

KANGAROO. The skin of the animal of that name, which makes a splendid leather, of firm texture. It is quite expensive, hence substitutes are on the market under the same name.

KID. A term applied to the shoe leather made from the skins of mature goats.

KIP. A term applied to leather made from hides weighing between fifteen and twenty-five pounds.

LACE STAY. A strip of leather reënforcing the eyelet holes.

LACE HOOK. An eyelet extended into a recurved hook, around which the lace is looped. It is most commonly used in men's and boys' shoes, although recently some have been invented for use in women's shoes with curved ends, to avoid catching the dress.

LACING. The operation of putting laces in shoes.

LAST. A wooden form over which the shoe is constructed, giving the shoe its distinctive shape.

LASTING. The process of making the uppers conform to the last in all respects. The operations of assembling and pulling over are parts of lasting.

LAYING CHANNEL. Turning down the lip or flap to cover the stitching.

LEVELING. Shaping the sole to the bottom of the last.

LIFT. The name given to one thickness of sole leather used in the heel. Top lift is the bottom lift, when the shoe is right side up, and is the last piece put on in manufacture.

LINING. The inside part of shoe, generally of cloth (dull) or sheepskin.

LINING CUTTING. The operation of cutting the cloth linings.

LINING-IN. The operation of putting lining inside of the shoe to cover insole or part of insole.

LOADING LEATHER. Filling the pores of the leather with glucose to increase its weight.

MAKING LININGS. Consists of closing up heel of lining; putting on top and side or eyelet stay.

MATCH MARKING. An operation performed on colored uppers, except black, to get different parts of the upper the same shade and color, and both shoes in the pair alike.

MAT. A term applied to a dull finish kid as distinguished from glazed.

MCKAY SEWED OR MCKAY. A shoe in which the outsole is attached to the insole and upper by a method named for the inventor.

MCKAY SEWING. Sewing through and through so that thread is seen inside of shoe.

MIDDLE SOLE. Any sole between out sole and insole.

MOCK WELT. McKay-sewed shoe with a double sole and having a leather sock lining. It is fair stitched to imitate a welt.

MONKEY SKIN. A. peculiar grained skin, and is considered in the trade as a fancy leather. It is often imitated.

MOROCCO. A name applied to leather originally made in Morocco. It is a sumac tanned goatskin, red in color, and is used in book binding. The name is also applied to a leather made in imitation of this, and to heavy, plump goatskins used for shoes.

MOLDING. Shaping the sole to fit the bottom of last.

MULES. The name applied to slippers with no counters or quarters.

NAP. The woolly side of hide, cloth, or felt.

NAUMKEAGING. Smoothing up the bottom with fine sandpaper. Sometimes the buffing grain.

NULLIFIER. A shoe with high vamp and quarter, dropping low at the sides, made with a short rubber goring for summer or house wear.

OAK TANNED. A process of tanning by means of a substance obtained from oak bark.

OIL LEATHER. Leather prepared by currying hides in oil. The hides are moist, that the oily matter may be gradually and thoroughly absorbed.

OOZE. A chrome tan calfskin treated on the flesh side in such a manner that the long fibers are loosened and form a nap surface; made in many colors.

OUTSIDE CUTTING. Cutting the leather parts of the shoe, as vamp, tip, top, etc.

OUTSIDE TAP. The tap used outside of men's or boys' heavy shoes.

OUTSOLE. The sole next the ground, on which all wear comes.

OXFORD. A low-cut shoe no higher than the instep lace, button, or goring, made in men's, women's, and children's sizes.

PACKER HIDES. Hides taken off in the large slaughterhouses. They are rated slightly higher in price, because great care and skill are used in taking them off.

PACKING. Placing a pair of shoes in a carton.

PACS. Coverings for the feet made of good quality calfskin, similar in form and appearance to the Indian moccasin. They do not have sole leather bottoms. If properly made, they are waterproof.

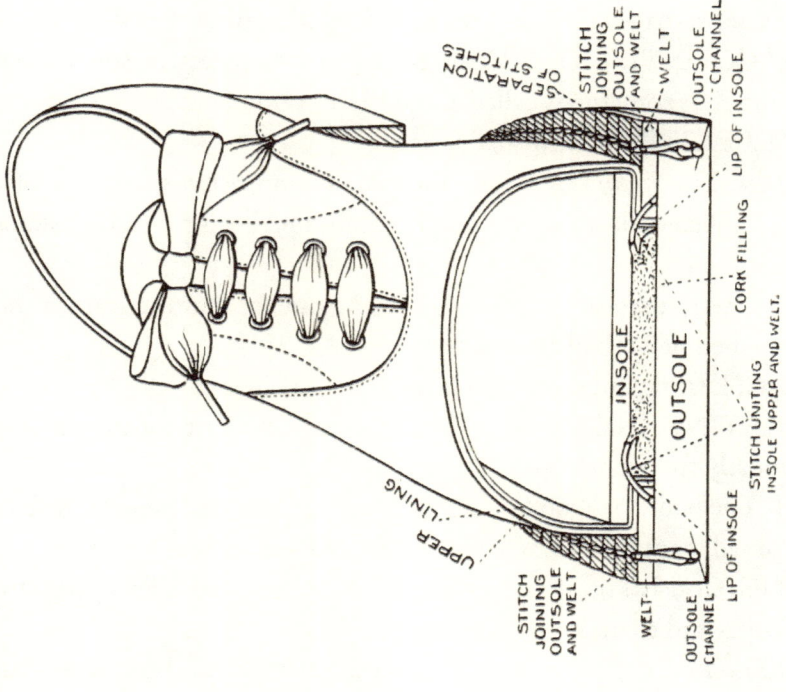

Cross Section of McKay Sewed Shoe.

Cross Section of Goodyear Welt Shoe.

PANCAKE. A term applied to one of the many artificial leathers formed from leather scraps, shaved thin, and cemented together under heavy pressure.

PASTED COUNTER. One that is cut from two pieces of sole leather pasted together. It is sometimes called a two-piece counter.

PATENT LEATHER. Varnished leather.

PATTERN. The model by which the pieces comprising the upper of a shoe are cut, applied collectively to upper as modified by the differing shape of these pieces.

PEBBLE. A term used in the process to bring out the grain of leather and give it a roughened or rubbed appearance.

PEGGING. Lasting out soles with pegs.

PERFORATING. Making very small holes around parts of upper. It is performed mostly for decoration.

POLISH. The name of ladies' or misses' front-lace shoe of higher cut than "bal," and named from Poland, where it originated.

PRESSING. Consists of a flat-press pressure for heels and soles, to prevent cracking of edges and to make parts adhere.

PORPOISE. This skin is sometimes used for leather and boot laces, but porpoise hides are ordinarily obtained from the white whale.

PULLING LASTS. Removing the lasts from shoes.

PULLING OVER. Pulling upper on the last and tacking it in position.

PUMP. A low-cut shoe originally having no fastenings, such as laces or buttons. A pump is cut lower than the instep.

PUMP SOLE. An extra-light single sole, running clear through to the back of the heel. A pump sole in former years was distinguished by its flexibility and was hand turned.

PUTTING ON TAP. Sticking half sole to the outsole.

QUARTER. The rear part of upper when a full vamp is not used. Term is used mostly in women's, and Oxfords or low shoes.

RAND. Made of sole leather about as wide as a welt, but thin at one edge. It is tacked to the heel so as to balance the heel evenly on the sole and fill any open space around the edge between sole and heel.

RAPID STITCHING. Sewing the sole to welt.

RELASTING. Consists in putting lasts in shoes from which the original lasts have been removed.

REPAIRING. A term applied to filling slight cracks in patent tips or patent leather.

ROAN. Sheepskin tanned with sumac. The process is similar in its details to that employed for morocco leather, but lacks the graining given to the morocco by the grooved rollers in the finishing. It imitates ungrained morocco.

ROLLING. The process of passing leather between rolls to make it firm and hard. Rolling consists in polishing the bottom on roll and brush.

ROUGH ROUNDING. Rounding outsole to the shape of last, and cutting channel in the welt-channeled shoes.

ROYALTIES. Sums paid for the use of machines to machine companies.

RUSSET CALF. Russet-colored calf is made from calfskins.

RUSSET GRAIN. Russet-colored grain is made from a split cowhide.

SABOT. The name of a one-piece wooden shoe, carved from a block of basswood. A novelty to Americans, but worn by people in the rural and manufacturing sections of Holland, Germany, and France.

SACK LINING. The lining inside the shoe and insole.

SANDAL. The name of a woman's strap slipper, or a sole worn by children. Originally fastened on the foot by straps.

SATIN CALF. A grain split, stuffed with oil, and smooth finished.

SCOURING BREAST. Sandpapering the front part of the heel.

SCREW-FASTENED. A shoe having the sole attached with screws, as in cheap or working shoes.

SEAL GRAIN. Usually a flesh split, with an artificial grain which is stamped or printed on the finished leather.

SECOND LASTING. The same as relasting. Term used most in turn work.

SHANK. The middle position of the bottom of the foot. Shank supports are placed in shoes to stiffen that part of the bottom. They are of steel, of wood, or of a combination of leather board and steel, and can be placed in the shoe any time before the out sole is laid.

SHANK BURNISHING. Polishing a black shank with hot iron.

SHANK FINISHING. Finishing the shank with blacking or in colors. The top lift is generally finished at the same time.

SHANKING OUT. Means making the edge of the shank thinner than the other part of the sole, and making it smooth.

SHEEPSKINS. Used largely for linings and for cheap shoes for women and children. It is too soft and weak in texture for heavy wear, and liable to split and tear.

SHORT VAMP. A foreshortened vamp. The distance between the extreme tip and the throat of the vamp shortened for appearances.

SIDES. Leather made from hides which are split into two sides down the back.

SIDE LASTING. Lasting the side of the shoe only.

SIZE. Shoes are measured by the length and width. The length is expressed by numbers and the widths by letters.

SKINS. A term used to represent the skin covering of small animals, such as goats.

SKIRTING. The outer parts of leather (hide), such as shanks, bellies, necks, etc.

SKIVING. Making the sole the same thickness in all parts. Skiving means cutting or shaving down to a thin edge. This operation may be done in the cutting department or stitching department.

SLIP. The name applied to spring heels or to soles. Slip is a thin piece of sole leather inserted above the outer sole.

SLUGGING. Driving slugs in heels, on part or all of the heel.

SOCK LINING. The lining for insole, inside of shoe.

SOFT TIP. A term applied to a shoe on which no boxing is used under the tip.

SOLES AND SOLE LEATHER. Name applied to pieces of leather of various thickness on the bottom of a shoe, usually made from heavy hides of leather. There are many varieties of soles: a "full-double" sole has two thicknesses of leather extending clear back to the heel; "half-double" sole is a full outer sole, with slip extending back to shank; single sole is self-defining; "tap" is a half sole.

SOLE LAYING. Sole laying is the operation of laying the outsole.

SORTING. The process of selecting and sorting soles, so that they may be put up in different qualities.

SPEWING. Shoes in stock sometimes become coated with a grayish white, powdery substance, that looks like mildew. This formation on leather that is not fully seasoned is called spewing, and the deposit is called bloom. It can readily be wiped off, and does not indicate any serious defect or trouble with the leather. It is not a mildew or growth, but apparently an exudation of materials used in tanning.

SPLITS. A name applied to split leather, that is, two or more parts of the hide.

SPRING HEEL. Consists of one or more lifts used between the outsole and upper. It is seen mostly in children's shoes and is often called wedge heel. It can also be put on outside instead of under the outsole.

STAMPING. The operation of putting size and width on the inside of the shoe. Parts of the uppers are often stamped or marked so that the whole are put together properly in the stitching room.

STAY. The name given to any piece of leather put in the upper to strengthen it or to strengthen a seam.

STAMPING BOTTOMS. The operation of stamping name on bottom. It is often performed in finishing rooms.

STAMPING CARTON. Putting the size, width, and other marks on carton.

STAMPING SIZES. Stamping sizes on heel part of the sole.

STANDARD-FASTENED. Nailing bottom on standard screw machine.

STAYING. Putting on a stay, generally heel stay.

STITCH SEPARATING. Marking between stitches so as to make them show to good advantage.

STITCH DOWN. A term applied to a flexible shoe used in the army, in which the top is turned out instead of under and stitched through the sole.

STITCHED ALOFT. A term used to indicate that the sewing stitches show on the bottom. No channel is necessary in this sole. It may be a slight groove. In stitching, the shoe is held bottom up, therefore the name "stitched aloft."

STRAIGHT LAST. One that is neither right nor left, and a shoe made over such a last can be worn on either foot. This term is sometimes applied to right and left shoes that have a barely perceptible outside swing.

STRIPPING. Consists of cutting in strips wide enough to cut soles all of equal size in length.

SUEDE. A trade term applied to kid skins, finished on the flesh side.

SWING. A term applied to the curve of the outer edge of a sole.

TACKING ON. Consists in laying the outsole on McKay's lasted shoes.

TACK PULLING AND TRIMMING OUT. Consist of preparing bottom for welting. It also makes it better for the operation.

TAMPICO. A variety of goat skins coming from the province of Tampico, Central America.

TAP. Half of a complete sole, often called half sole when used under outsole.

TAN. Tan is a sort of brownish leather.

TANNING. Tanning is the process of converting hides or skins into leather.

TAP TRIMMING. Shaping the tap to conform to the sole.

TAWING. The process of making leather by soaking hides in a solution of salt and alum, or by packing them down with dry salt and powdered alum. Used to prepare skin rugs and furs.

TEMPERING. The operation of wetting the leather in water to take hardness out and make leather "mull," so that it may be worked easier.

TIP. The toe piece which is stitched to the vamp and outside of it. Stock tip is a tip of the same material as the vamp. Patent tip is a patent leather tip. Diamond tip refers to the shape extending back to a point. Imitation tip-stitching across the vamp is imitation of a tip.

TIP CUTTING. Cutting the tip which goes on the toe of the vamp.

TOE AND HEEL LASTING. Lasting heel and toe.

TOE PIECE. A piece attached to cut-off vamp to lengthen it.

TONGUE. A narrow strip of leather necessary on all laced shoes.

TOP. The part of the upper above the vamp; tip of shoe.

TOP CUTTING. Cutting the top only.

TOP FACING. The strip of leather or band of cloth around the top of the shoe on the inside is called the top facing. It adds to the finish of the lining, and is sometimes used to advertise the name of manufacturers by a design of letters woven or sewed on it.

TOP LIFT. The lift which is next to the ground.

TOP LIFT SCOURING. Sandpapering top lift of heel to make it smooth.

TOP STITCHING. Consists of stitching across the top and down the side.

TREEING. Shaping the shoe, making it smooth. Produces the same effect as ironing, although no hot iron is used. It makes the upper plump and gives it a good finish and "feel."

TRIMMING CUTTING. Cutting stays, facings, and other small parts of the upper.

TRIMMING VAMP. Cutting off hanging or surplus thread.

TURNING. To turn shoe right side out. Also turning upper right side out.

TURNED SHOE. A lady's fine shoe that is made wrong side out, then turned right side out, which operation necessitates the use of a thin, flexible sole of good quality. The sole is fastened to the last, the upper is lasted over its wrong side out, then the two are sewed together, the thread catching through a channel cut in the edge of the sole. The seam does not come through to the bottom of the sole where it would chafe the foot on inside.

UPPER. A term applied collectively to the upper parts of a shoe.

UNGRAINED. Smooth surface.

VAMP. The lower or front part of the upper of a shoe. It is the most important piece of the upper and should be cut from the strongest and cleanest part of the skin. "Cut-off" vamp is one that extends only to the tip, instead of being continued to the toe and lasted under with the tip. Whole vamp is one that extends to the heel without a seam.

VAMPING. Stitching the vamp to the top.

VAMP CUTTING. Cutting vamp with or without the tip.

VELOUR. A finish for calf leather. It is the French name for velvet and is used in the shoe trade for a patent chrome-tanned calf leather. It is, an excellent leather and has a smooth and velvety finish.

VELLUM. A name for skins that are made into a variety of parchment.

VENEERING. Consists in making soles, whole or part, heavier, by means of leather board or other material fastened to the sole by an adhesive.

VESTING. A material originally designed for making vests. As used in shoes, it is made with fancy-figured weave, having a backing of stiff buckram or rubber-treated tissue to strengthen it.

VISCOLIZING. A patent method of water-proofing sole leather by the use of partly emulsified oils with a water-resisting tendency. Viscolized soles are used in hunting and sporting boots.

VICI. A patent trade name for a brand of chrome-tanned kid.

WASH LEATHER. An inferior quality of chamois.

WELT. A narrow strip of leather that is sewed to the upper of a shoe with an insole leaving the edge of the welt extending outward, so that the outsole can be attached by sewing through both welt and outsole, around the outside of the shoe. The attaching of the sole and upper thus involves two sewings, first the insole, welt and upper, then the outsole to the welt. The name is applied to the shoe itself when made in this way to distinguish it from a turned, or McKay sewed shoe. This is the method used by cobblers in the production of hand-sewed shoes to fasten the sole and upper together. Goodyear welt is a welt in which the sewing is done by a machine named for the inventor. There are very few hand-welted shoes made.

WELT BEATING. The flattening out of the welt, making it smooth.

WELTING. Sewing the welt to shoe.

WHITE ALUM. Bleached leather tawed with white alum.

WOODEN CASE. Large box for twelve or more pairs.

# CHAPTER TEN.

## LEATHER PRODUCTS MANUFACTURE

THE use of gloves is so old that relics of them have been found in the habitations of the cave dwellers. The Romans used them as decorative articles of dress and the Greeks to protect the hands when doing heavy work.

The gloves of ladies and gentlemen in the days of Queen Elizabeth, and before and after, were most beautiful in hand workmanship and embellishments, but they were usually shapeless things, and in these days no one would wear them; they are not to be compared with the elegant style and artistic finish of the modern product.

When the social world was restricted, so to speak, in the numbers of its members who could afford some of life's luxuries, the use of the glove was confined largely to royalty, nobility, and the well-to-do. And the trade not being extensive, prices were high—being added to by decorative elaboration in needlework in order that the manufacturer and his employees might extract as much money as possible from the ultimate buyer. While glove making is now one of the stabilities of modern manufacture, it is, nevertheless, constantly changing in styles, due to eagerness for novelties and new fashions.

Glove making of leather, in a rough, crude form, was carried on in this country to a very limited extent in New York State as early as 1760, by glove makers brought from Scotland to settle on the grants of Sir William Johnson, in Fulton county. But there was no general market for the home product until one was found in Albany in 1825. These early gloves, crude and clumsy, were cut with shears from leather by means of pasteboard patterns, and men did the cutting and women the sewing. Dies were later introduced, and this led to a great improvement in the character of the output.

But a still greater step forward was taken when the sewing machine was introduced in 1852. This abolished handwork entirely, but still the industry remained largely of a domestic nature, since it could be carried on at home with a machine as well as in a factory. Later steam power was installed in factories with which to run the machines. The cutting of gloves, and the stitching on the backs, was done before the gloves were sent out to be completed in workers' homes.

As in everything wherein power can be substituted for hand labor in these days, the methods of glove manufacture have undergone a great transformation. The treating of skins in a great tub, three feet deep, whole dyeing and scouring, in rooms of high temperature, has been displaced by putting skins and colors into a cube shaped box, which, revolving with an irregular motion, produces the same results more quickly than by the primitive way. But when color is to be applied to but one side the process is the same as of old,—hand use of a brush while the skin is stretched out on a slab.

When taken from the stock on hand to be made into gloves, the first thing done to skins by some glove makers is to "feed" them with eggs—not eggs of suspicious merits, but good enough for table use. And of these nothing is used but the yolk. One glove maker imports from China large quantities of the yolks of duck eggs for his work, and his yearly consumption of yolks amounts to seventeen thousand.

When the skins leave the dyehouse, they are rapidly dried in steam-heated lofts; and while stiff and rough they are, or were, worked into softness and smoothness over a wooden upright standard, called a stake, at the top of which is fitted a blunt semicircular knife. Over this the skin is drawn by hand, back and forth; until it becomes as pliable and delicate as silk. When this work was done manually it was most laborious. But now it has been mostly taken over by very ingenious machinery, which looks, in operation, as if it would tear a skin into fragments by the way it snaps and pulls at it, but which is adjustable to such nicety of action and power that the work is done exactly as it is wanted.

The next operation is to pare the skins to uniformity of thickness. This also was handwork for a long time, done with a peculiarly shaped knife, but now emery-coated wheels, with rounded edges, are used by the workers, who, with their aid, do just as good and much faster work in drawing and thinning the skins with absolute precision. This completes the treatment of the skin.

Now the function of the cutter begins, and he must be a workman of experience and good judgment, in that he must contend with the inconstant inelasticity of the skin, reducing it to uniform resistance. He must get so many pieces of glove size from each skin, and suit the pieces to particular features of the skin. When done with a skin he must have left, as useless, only trifling strips and shreds. The shapeliness of the glove which a woman draws over her hand, depends altogether upon the intelligence and skill of the cutter. In American factories the cutter is usually from some glove-making center in Europe and from a family whose occupation has been glove making for centuries.

A punch next cuts these glove pieces into shape, forming and dividing the fingers, slitting the buttonholes, providing side pieces for fingers and thumbs, and also the fragments used for strengthening the buttonholes. The sewing, formerly the handiwork of women, is now done on machines of capacity for exceptionally fine quality of intricate stitching. The number of glove sizes made is sufficient to meet every likely demand. When sewn, and the buttons or fastenings put on, they pass beneath the critical eye of an inspector for scrutiny as to faults. Then they are finally shaped on a hot metal hand, smoothed, banded, boxed, and sent to the salesroom for shipment.

The first and fourth fingers of a glove are completed by gussets, or strips, sewed only on the inner side; but the second and third fingers require gussets on both sides to complete the fingers. In addition to these, small, diamond-shaped pieces are sewed in at the roots of the fingers. Special care is necessary in sewing in the thumb pieces, as poorly made gloves usually give way at this point.

Natural lined gloves are now common enough, although it is not many years since they were regarded as impracticable. These are made from pelts of various animals with the hair left on the skin to form the lining.

### Automobile and Furniture Leather

For automobile and furniture leather only choice hides should be used. The kind of hides generally employed for this class of leather are French and Swiss, as these run full and plump on the bellies, are free from cuts on the flesh and are of clear grain. The hides are trimmed before placing them in the soaking pits, all useless parts, such as nose, shanks, etc., being cut away.

After remaining in soak for a day or two, the hides are hauled out, fleshed, and returned to the soaks for thorough softening. When thoroughly soaked, they are toggled and reeled into the first lime. The first lime must be a weak, mellow lime, or a harsh grain will show after the leather is tanned. The hides are reeled over into stronger limes every day for seven days, when they are ready for unhairing. After coming from the limes, the hides should go into a pit of soft water heated to about ninety degrees Fahrenheit and left over night before starting in to unhair. After unhairing, they are thrown into a vat of clean water and thoroughly worked out on the grain to remove short hairs and send and are then ready for bating. One that has a little bacterial action is preferred to an acid bate. After bating, the hides are given a good scudding on the grain and are then ready for the tanning liquors.

The liquors are made of hemlock and oak and are used very weak on the start. The hides are suspended for a day in a liquor not over six degrees specific gravity reading in strength, and the following day shifted over into a stronger liquor. The stock is given stronger liquors every day until tanned enough for splitting.

The stock is struck out smoothly and brought to the machine for splitting. The buffing is first taken away and sold for hat bands, pocket-books, etc. The grains are finished and the splits are returned to the tanning liquors to be thoroughly tanned. As soon as the splits are tanned, they are washed up, drained, and then drummed in the drum in a sumac liquor. They are now scoured, and, after being well set out, are given a good oiling with cod oil.

They are now tacked out on the frames and dried out. They are next taken from the frames and boarded by hand over the table. The splits are taken to the japan shop and are tacked out again and are ready for the first coat of daub. Two coats are applied. After each coat, the splits are well rubbed down, when they receive the slicker coat. The color coats are now applied, and after drying out, the leather is grained up and finished.

# CHAPTER ELEVEN.

## RUBBER SHOE MANUFACTURE

EXAMINE the rubbers we wear during the winter and stormy weather.

Rubber shoe coverings are made to protect the shoe from water and snow and may be in the form of either slippers or arctics. The covering is rendered waterproof by means of a compound rubber.

Rubber is the name given to a coagulated milky juice obtained from many different trees, vines, and shrubs that grow on that part of the earth's surface which forms a band some three or four hundred miles on either side of the equator.

Rubber is graded commercially, according to the district where it is found. In the order of importance it may be divided into three general sorts, viz., American, African, and Asiatic. The best and largest quantities of rubber come from Brazil, along the banks of the Amazon River. The countries in the northern and western part of South America, and the Central American States and Mexico furnish considerable rubber. Eastern and western Africa also produce many kinds of rubber in large quantities, though somewhat inferior to the Brazilian product. The Asiatic rubbers are unimportant in quantity, and, excepting the rubber obtained from cultivated trees in Ceylon, are decidedly inferior in quality.

The fluid rubber obtained from Brazil is called Para and is used principally in the manufacture of rubber footwear. The method of gathering and coagulating the rubber juice (called latex) varies in the different countries. The native first clears a space under a number of trees and proceeds to tap the trees with a short-handled ax, having a small blade, by cutting gashes in the bark. A cup is fixed under each cut to catch the fluid as it flows out. As fast as the cups are filled, they are emptied into a large vessel and carried to the camp to be coagulated. A fire is started in a shallow hole in the ground, and palm nuts, which make a dense smoke, are thrown on. An earthen cover which has a small opening on top is placed over the fire, allowing the smoke to escape through the opening. A wooden paddle is first dipped in clay water and then into the latex and then held over the smoke. The heat coagulates a thin layer of rubber on the paddle. It is dipped again and again in the latex and smoked each time. After being dipped many times, a lump (called biscuit) of rubber is formed. A cut is made in the biscuit and the paddle removed. Then the rubber is ready for market. The world's crop of rubber in 1911 was about ninety thousand tons.

Crude Rubber.

Few people realize the number of operations necessary to produce from the crude biscuit of India rubber the highly finished rubber shoe of to-day. Briefly stated, the various steps are washing, drying, compounding, calendering, cutting the various parts, making or putting these parts together, varnishing, vulcanizing, and packing. Each of these processes requires a distinct and separate department, and many of these processes are subdivided into minor operations.

Washing and Drying.

The huge stock of Para rubber, that is rubber obtained from the Amazon section, to be found in any of the leading rubber factories counts well up into the thousands of dollars. With rubber at or near $1.50 per pound, a stock of ten to fifty tons runs up into the five or six figures.

This crude rubber, as it comes from the Amazon, contains more or less dirt, pebbles, and other foreign substances, which must be removed.

The large cakes of crude rubber are first broken up by a cracker machine, consisting of two large, revolving steel cylinders, from which the product falls into pans or trays. It goes then to a machine known as a "washer" or "sheeter," where it is run between revolving cylinders, upon which a continuous spray of clean water is maintained. After being rolled into rough sheets, it is put into a tank, from which it is taken to the "beater" machine, in which water runs continuously, and then it is washed again and "sheeted out." It is then dried in one of two ways.

(1) The older way. The sheets are hung over rods in a large room, and allowed to dry in the air. To facilitate the same, a fan or blower is often used to cause a circulation and removal of the moisture-laden air. This requires a period of from one to two or three months.

(2) The second method is called vacuum drying. This process is gradually being introduced, so that now probably more rubber is dried in vacuum than by air. The vacuum drier consists of a large iron cylinder filled with plates, through which steam is allowed to circulate. The rubber is placed on the plates and the air is exhausted from the cylinder by means of an air pump until very nearly twenty-six degrees of vacuum are obtained. By this process only from two to three hours are required to produce perfectly dry rubber.

The making of a rubber shoe is not the simple matter which might at first be supposed. An ordinary rubber shoe consists of at least seven or eight different parts, sometimes twenty-one parts to a pair, while a high-button gaiter has seventeen distinct parts, and a rubber boot has twenty-three different pieces. There are insoles, outsoles, stays, piping, foxing, and a dozen other different pieces, each one of which is necessary to the proper construction of a rubber shoe or boot. The thinner sheets for the uppers are cut by hand, the deft work of the cutters in following the patterns outlined on the sheets being the result of years of practice. The sheets of rubber from which the uppers and soles are cut are at this stage of the work plastic and very sticky. It is necessary on this account to cut the various pieces one by one, and keep them separate. The soles and some of the heavier pieces are dried out by the machine, and the heels are made by a special machine, but by far the greater part is done by wonderfully skilled hands. All of these parts which go to make a shoe, or the twenty-three parts which go into a boot, are collected and sent to the making department, which, in most factories, is a large room containing several hundred operatives, each working by herself, and bringing the many separate parts into the fully finished footwear.

The sheets of rubber, after being dried, are taken to the "compound" room, where they are sprinkled with whiting, to prevent sticking, and weighed. Next they are taken into the calender room to a "mixer," by means of which the rubber is combined with other substances, which include sulphur, litharge, whiting, lampblack, tar, resin, lime, palm oil, and linseed oil.

Calender Room.

There are different calendering machines. The ones called the upper calenders form sheets of rubber stock for the upperpart of the shoe. The soling calenders form the stock for the sole or bottom part of the shoe; other calender machines are used to coat a layer of gum on one side of the fabrics used for lining and various strips, fillers, toe, and heel pieces. The gum sheets are sent to the cutting room.

Generally, linings for nine pairs of shoes are cut at once. The linings are cut both by hand and by machine. Men who cut with dies, by hand, stand at the bench and use iron mallets, like those used in cutting heels. Inner soles, heel pieces, and linings are all cut by means of dies in the same manner.

The edges of the several parts are spread with cement, and then the parts are taken to the making room and distributed. In the making department the boots and shoes are put together. Women make the light overshoes; men make the heavy ones. Rubbers are made by women, but men put on the outer soles.

Cutting Room.

Linings are first applied smoothly to a wooden last and cemented together, the cement side out. The rubber parts are then stuck on and rolled firmly with a small hand roller. Young women become very skilled in this work, taking up the several parts in rapid succession, placing them accurately upon the last, and rolling and pounding them firmly together.

Perhaps the most interesting single process is that of putting the rubber boot together. This work is done by men, and requires, in addition to accurate eyesight, rapid and very deft movements of the hand and considerable strength. No nails, tacks, or stitching are required. The natural adhesiveness of the rubber, assisted by the use of rubber cement, holds the parts solidly together.

In the making of the shoe the last is covered with the various pieces which are so made as to adhere where they are placed. It is exact and nice work fitting all these pieces perfectly, each edge overlapping just so far and no farther. The lighter shoes are made by women, but the heavy lumbermen's shoes, arctics, and especially the boots, are made by men, for this work needs strength as well as dexterity.

The goods which require varnishing are put on racks and treated with a mixture of boiled linseed oil, naphtha, and other materials, which are applied with brushes, and impart a gloss to the surface.

On vulcanizing boots and shoes, the shoes are placed on racks supported by iron cars, which are run over tracks into the vulcanizing chamber. This consists principally of a large room provided with a steam coil on the floor. The temperature rarely exceeds two hundred and sixty degrees Fahrenheit. In vulcanizing shoes, the heat is increased gradually from the beginning, about one hundred and eighty degrees Fahrenheit, otherwise the goods would be blistered, due to the rapid evaporation of moisture and other volatile constituents. They are kept in these heaters from six to seven hours. This causes a union of sulphur and rubber, which is not affected by heat or cold.

They are wheeled on another truck to the packing room, where they are inspected, taken from the lasts, tied together in pairs, or placed in cartons, as the case may be. They are then sent to the shipping room to be packed in cases ready to be delivered to the cars waiting at a side track of the railroad, or sent to the storehouse until they shall be called for by the jobbers or retail dealers.

An important branch of the rubber business is the manufacture of tennis shoes. This is a generic term, which is applied to all kinds of footwear having cloth tops and rubber soles. As the name indicates, they were first used in playing the game of tennis, but they have come into very general use as warm weather and vacation shoes, and every year shows an increased popularity. These shoes are made in a similar manner to the rubber shoes, the rubber soles being cemented to the cloth uppers and vulcanized the same as the rubber over-shoes. Many different styles are made, and each year shows some improvements in the shapes, in the textiles which are used, in the colors and combinations of soles and uppers.

Rubber shoes should not be expected to give satisfactory service unless properly fitted. If too short, too narrow, or if worn over leathers with extra heavy taps, or unusually thick, wide soles, strains will be brought upon parts not designed to stand them and the rubber will give way. Rubber goods, particularly boots, if too large will wrinkle and a continued wrinkling and bending is liable to cause cracking.

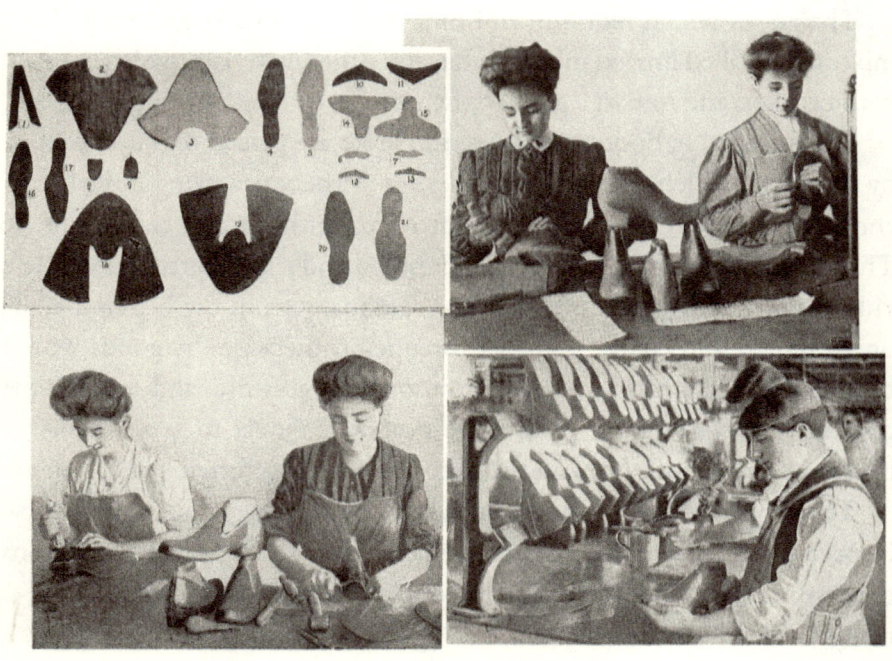

Putting together the Parts of a Rubber Shoe.

Extreme heat or cold should be avoided. Rubber boots or shoes should never be dried by placing them near a heater of any kind. If left near a stove, register, or radiator, the rubber is liable to dry and crack. If left out of doors in winter, or in an extremely cold place, they will freeze. Then when the warm foot is put into them and the rubbers are worn, the rubber will crack.

Oil, grease, milk, or blood will cause rubber to decay in a very short time. If spattered with any of these, the rubber should be promptly and thoroughly cleaned with warm water and soap.

The oil in leather tops will rot rubber, so that care should be taken in storing and packing to prevent the leather and rubber from coming in contact.

Various heavy goods are advertised as proof against snagging. It should be remembered, however, that no rubber can be made strong enough to be absolutely proof against tearing or puncturing by extremely sharp edges, such as stiff stubble, sharp-edged rocks, broken glass, etc.

Mud, barnyard dirt, or filth of any kind should never be allowed to dry on rubbers. They should be" cleaned as carefully as leather boots or shoes.

Exposure to strong sunlight for any length of time produces an effect on rubbers similar to that of putting them near a stove or radiator. Rubbers should not be left in the sun to dry. When not in use they should be kept in a cool, dark place.

## RUBBER HEELS

Rubber heels are generally made for boots and shoes as follows. The compounded rubber is sheeted on a calender roll, on a drum, until several layers are obtained, thus making a sheet of about one inch in thickness. The heel is cut out from this sheet by means of a die and placed in a mold. It is there subjected to an extremely high pressure, generally obtained by hydraulic power. The plates of the press are heated with live steam. The heels are removed at the end of nine or ten minutes and the sheet which was formerly nearly an inch in thickness is now only about half an inch and has by pressure been molded into the shape of the heel desired, is semi or partially vulcanized, and also is imprinted upon the bottom with the name or other brand of the company.

Heel-making Department.

The cup-shaped portion of the heel is now coated with a layer of rubber cement, and firmly placed on the boot ready to go to the vulcanizer, where vulcanizing of the heel is then completed.

Many articles of rubber are vulcanized by the use of chloride of sulphur, which process is sometimes known as "cold cure." The action of sulphur chloride itself is so violent that it must be diluted, and for this purpose carbon bisulfide is often used. In some cases, as, for example, the manufacture of tobacco pouches, the articles are submerged for from one to two minutes in the liquid, then removed and washed thoroughly. In another case, as in the manufacture of some kinds of rubber cloth, such as hospital sheeting, the coated cloth is suspended in a suitable room and the chloride of sulphur and carbon bisulfide mixed and evaporated by action of heat so that the cloth is subjected to the action of vapor alone. Only articles with comparatively thin walls can be successfully vulcanized by the cold cure, as at best the vulcanizing action of the chloride is only superficial.

No account of vulcanization processes as employed in the manufacture of rubber goods is complete without the mention of "steam cure." A great variety of rubber goods under the general term of mechanical sundries are cured by this method. This includes rubber matting, door mats, water bottles, druggists' sundries, etc. This process consists in brief of submitting the articles to be vulcanized to the action of live steam for from half an hour to an hour, or until the goods are thoroughly vulcanized. The temperature and duration of time required depend to a considerable extent upon the thickness of the walls of the article. In order to prevent the goods from being pitted and damaged by the action of steam, they are wrapped with cloth or imbedded in pans of soapstone. A great variety of rubber tubing is cured by this method.

In rubber cloth making, the crude rubber is put through the washing process, dried and mixed with sulphur, litharge, coloring matter, etc., and then is taken to the cement room, where it is "cut" with naphtha, forming a thick paste or dough. This is taken to the spreading room in large tubs and fed into the roller machine, which is like a long table made of steam pipes placed horizontally in a single layer. Below one end is a roll of cloth, which is passed between two iron rollers on the end. The dough is fed in between these rollers and is spread smoothly, over the cloth, which is rolled up and removed to a heating room, where it is unrolled and hung on racks, and then subjected to sufficient heat to cause the combination of the sulphur and rubber.

## Chemistry in the Manufacture of Rubber Goods

Too much stress cannot be laid upon the importance in all rubber factories of the chemical department. During the last two or three years there has been an unusual development along these lines, and to-day no factory for the manufacture of rubber goods is complete that does not possess a well-equipped laboratory. Not only does this department enable the manufacturer to control the purity and uniformity of his compounding ingredients and the innumerable grades of crude rubber, but, what is of even greater importance, it enables him to inaugurate research work as applied to his particular line of manufacture. This part of laboratory work is already producing results not only of scientific interest, but of very great practical and economic value. Still another rôle of the modern chemical laboratory is to exercise a control over the finished material, so that the manager of the works may be in possession daily of reasons for any variation detrimental to the standard of his products.

## Rubber Terms

ANKLE PIECE. A large piece of light sheeted gum, which goes around the ankle and extends about halfway up the leg.

BACK STAY. A piece of frictioned sheeting similar to the side stay in shape and placed at the back of the heel and ankle.

GUM COUNTER. A piece cut out of sheeted gum, on the under side of which is placed a counter form or a piece of frictioned sheeting.

OUTER FILLER. A filling sole cut from rag-coated or frictioned sheeting, and designed to fill up the hollow on the bottom caused by bringing the edges of the gum vamp and counter underneath.

INNER SOLE. Usually made of felt or sheeting coated on one side with rag stock. In lasting up, the bottom edges of the lining (which have previously been cemented) are pulled under and adhere to the inner sole.

LEG COVER. A piece of sheeted gum rolled upon a piece of frictioned sheeting called the leg form.

LEG LINING. The lining, usually of felt or wool netting, for the leg.

PARA. A name given to rubber from Brazil.

PIPING. Strips of frictioned sheeting used to join the lining together over the instep and up the back, and also to hold the lining up on the tree by passing a strip over the top.

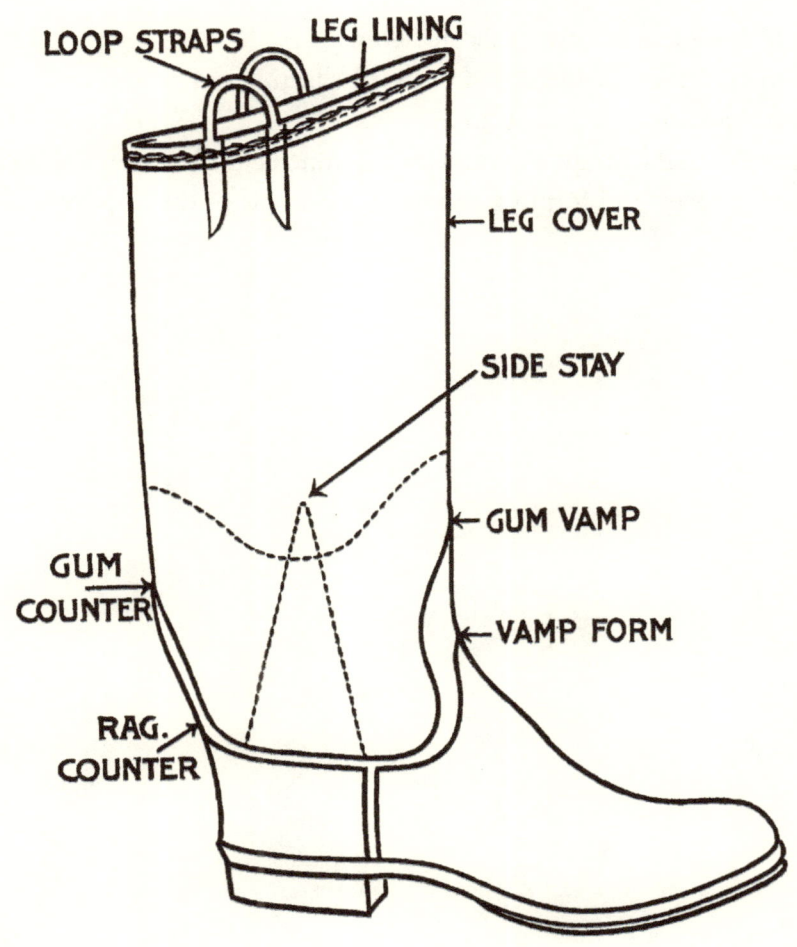

Parts of a Rubber Boot.

RAG COUNTER. Quarter stiff is a counter piece cut out of rag-coated or frictioned sheeting, which gives stiffness to the counter.

SIDE STAY. A spike-shaped piece of frictioned sheeting, placed on each side of the ankle.

RAG SOLE. A sole stiffening cut out of a sheet of rag stock, which covers the whole bottom. The edges are skived to make a perfect edge.

TOE FILLER. A rag-stock filling sole to fill up the hollow on the bottom caused by attaching the lining to the inner sole.

TOE LINING. The lining for the vamp, of the same material as the leg lining.

VAMP. A piece cut out of sheeted gum.

VAMP FORM. A piece of frictioned sheeting cut to the shape of the vamp, and put on over the toe lining.

WEB STRAPS. Straps put on with the joined ends between the leg lining and the leg cover, and forming a loop on the inside of the boot to pull it on with.

# CHAPTER TWELVE.

## HISTORY OF FOOTWEAR

WE find that primitive footwear, in common with all other beginnings, was of the crudest nature and took the form of the simple sandal. It is probable that man first protected his foot from the rough way by simple pieces of hide, which were bound to the bottom of the foot. The sandal, among the most primitive, is the type of footwear worn to-day. The sandal was simply bound to the foot by thongs of hides, which were brought between the toes and tied around the ankle.

At about the Elizabethan period, shoe making had really become a very fine art. Some foot creations were made by the Court shoemakers that reflected the individual taste of the monarch, and so great was the competition to produce something novel that very often the styles assumed a grotesque aspect. The toes were elongated so that sometimes they were carried up and fastened by cords and tassels to the tops of the shoes, and it finally became necessary to enact a law to prevent such outrageous types of footwear. The slippers of this period were of the extremely high-heeled variety, and small fortunes were often spent on their ornamentation. They were mostly of the turn-shoe type, and samples which are preserved show the excellent workmanship that was in vogue at that time.

We now come to the first shoemaker in America. When the *Mayflower* made the second trip to America, she carried among others a shoemaker named Thomas Beard, who brought with him a supply of hides. Seven years afterwards there arrived one Phillip Kertland, a native of Buckinghamshire, who settled in Lynn in 1636.

Kertland was the pioneer shoemaker of Lynn and for years he successfully worked at his craft, teaching others his methods and ways, so that fifteen years after his arrival, Lynn was not only supplying the requirements of its inhabitants, but was also sending a part of its products to the port of Boston.

As early as 1648 we find tanning and shoemaking mentioned as an industry of the colony of Virginia, special mention being made of the fact that a planter named Matthews employed eight shoemakers on his premises. Legal restraint was placed on the cordwainer in Connecticut in 1656, and in Rhode Island in 1706, while in New York the business of tanning and shoemaking is known to have been firmly established previous to the capitulation of the Province to England in 1664. In 1698 the industry was carried on profitably in Philadelphia, and in 1721 the Colonial Legislature of Pennsylvania passed an act regulating the material and the prices of the boot and shoe industry.

Prior to 1815 most of the shoes were hand sewed, a few having been copper nailed. The heavier shoes were welted and the lighter ones turned. This method of manufacture was changed, about the year 1815, by the adoption of the wooden shoe peg, which was invented in 1811 and soon came into general use. Up to this time little or no progress had been made in the methods of manufacture. The shoemaker sat on his bench, and with scarcely any other instrument than a hammer, knife, and wooden shoulder stick, cut, stitched, hammered, and sewed until the shoe was completed. Previous to the year 1845, which marked the first successful application of machinery to American shoemaking, this industry was in the strictest sense a hand process, and the young man who chose it for his vocation was apprenticed for seven years, during which time he was taught every detail of the art. He was instructed in the preparation of the insole and outsole, depending almost entirely upon his eye for the proper proportions; taught to prepare pegs and drive them, for the pegged shoe was the common type of footwear in the first half of the last century; and familiarized himself with the making of turned and welt shoes, which have always been considered the highest types of shoemaking, as they require exceptional skill of the artisan in channeling the insole and outsole by hand, rounding the sole, sewing the welt, and stitching the outsole. After having served his apprenticeship, it was the custom for the full-fledged shoemaker to start on what was known as "whipping the cat," which meant traveling from town to town, living with a family while making a year's supply of shoes for each member, then moving on to fill engagements previously made.

The change from which has been evolved our present factory system began in the latter part of the 18th century, when a system of sizes had been drafted, and shoemakers more enterprising than their fellows gathered about them groups of workmen, and took upon themselves the dignity of manufacturers.

It was soon found that the master workman could largely increase his income by employing other men to do the work while he directed their efforts, and this gradually led to a division of the labor: the shoe uppers, which had prior to this time been sewed by men using waxed thread with bristles, now were done by women, who often took the work home.

One workman cut the leather, others sewed the uppers, and still others fastened uppers to soles, each workman handling only one part in the process of manufacture.

We find that in the year 1795 the evolution of the factory system had reached a stage where in Lynn alone there were two hundred master workmen, employing six hundred journeymen and turning out three hundred thousand pairs of shoes per year. The entire shoe was then made under one roof, and generally from leather that was tanned on the premises.

Factory buildings were not at this time of a very pretentious nature and did not by any means represent the amount of work undertaken by the proprietor; for the small ten by ten factories, which are even to-day in existence in some of the backyards of Lynn homes, came into existence at this time. Many farmers found that shoemaking was a remunerative occupation in the winter, and they, and perhaps their neighbors, gathered in these shops and took from the different factories shoes on which to fasten the soles, or uppers to bind, which, after completion of the work, were returned to the factory, where they were finished and sent to market packed in wooden boxes. It was in this way that the industry prospered and developed up to the period of the introduction of machines, which happened but a little over half a century ago.

Up to the year 1811 absolutely no machinery was used in the making of shoes. This year shoe pegs were invented and a machine for making them. The pegged shoe became very widely worn, but it was not until 1835 that any machine for driving pegs was made, and even at this time the machine was but an indifferent success. It was a hand machine and its work was by no means of a reliable nature.

The first machine to be widely accepted by the trade was the "rolling machine." This was used for rolling the sole leather under pressure, and it is said that a man could perform in a minute with this machine the same office that he would have required half an hour to have performed with the old-fashioned lapstone and hammer. This was followed in 1848 by the most important invention, the "sewing machine," which was perfected by Elias Howe, and was soon followed by a machine which sewed with waxed thread and made it possible to sew the uppers of shoes in a much more rapid, reliable, and satisfactory manner than had ever been done by hand. This, too, was soon followed by a machine which split the sole leather and by another for buffing or removing the grain.

In 1855 William F. Trowbridge, who was a partner in the firm of F. Brigham & Company, of Feltonville, Massachusetts, then a part of Marlboro, conceived the idea of driving by horse power the machines then in use. The introduction of power became very general, so that in the year 1860 there were scarcely any factories which were not driven by either steam or water power.

The year 1858 was marked by the invention by Lyman R. Blake of the McKay sewing machine, which probably more than any other has exerted a revolutionary effect on the industry.

The McKay machine did not at this time sew the toe or heel; the sewing was started at the shank and carried forward to a point near the toe on one side, and the same operation repeated on the other side; but it seemed to possess great possibilities and created a great deal of interest throughout the trade. It was, of course, a very crude machine and very different from the McKay machine of to-day. It was set on a bench and the shoe to be sewed was placed over a horn, and the sewing was done from the channel in the outsole through the sole and insole. Colonel McKay immediately started to improve the machine. He employed skilled mechanics to work on it and attempted to introduce it in different factories, but encountered a great deal of opposition and criticism in regard to its future. It is said that he offered to dispose of the machine to the shoemakers of Lynn and allow them its exclusive use if they would pay him three hundred thousand dollars, an offer which was not accepted.

The machine left a loop stitch and a ridge of thread on the inside of the shoe, but it filled the great demand that existed for sewed shoes, and many hundreds of millions of pairs have been made by its use.

While Colonel McKay had met rebuff and discouragement in attempting to introduce his machine, the public necessity was such that manufacturers were obliged to take it up immediately; but Colonel McKay was still embarrassed by lack of capital to carry on his rapidly increasing business. It was at this time that a system of placing machines in factories, which system has proven to be the most potent factor in the upbuilding of the shoe industry, was started. This was a royalty system, whereby the machine or machine owner participated in the profits accruing from the use of the machine.

It hardly seems that there can be any question as to the principle of royalty being one of the greatest forces in building up the successful industry which we have to-day; it afforded an easy means whereby machines could be introduced without entailing hardships on the manufacturers, who, had they been obliged to pay the actual worth of the machines, would have been entirely unable to adopt them. Instances are known where hundreds of thousands of dollars were spent on machines, which machines were abandoned without having made a single shoe.

At the time of the introduction of the McKay machine, inventors were busy in other directions, and as a result, came the introduction of the "cable nailing machine." This was provided with a cable of nails, the head of one being joined to the point of another; these the machines cut into separate nails and drove automatically. At about this time also was introduced the "screw machine," which formed a screw from brass wire, forcing it into the leather and cutting it off automatically. This was the prototype of the "rapid standard screw machine," which is a comparatively recent invention, and is very widely used at the present time as a sole fastener on the heavier class of boots and shoes. Very soon thereafter the attention of the trade was attracted to the invention of a New York mechanic for the sewing of soles. The device was particularly intended for the making of turn shoes and afterwards became famous as the "Goodyear turn shoe machine."

Closely following the Goodyear invention came the introduction of the first machine used in connection with heeling,—a machine which compressed the heel and pricked holes for the nails; this was soon followed by a machine which automatically drove the nails, the heel having previously been put in place and held by the guides on the machine. Other improvements in heeling machines followed with considerable rapidity, and a machine came into use shortly afterwards which not only nailed the heel, but which was also provided with a hand trimmer, which the operator swung round the heel, after nailing. From these have been evolved the heeling machines in use at the present time.

One of the early uses to which the sewing machine was put was the sewing together of the pieces of soft and pliable leather which make the upper of a shoe—a simple thing, involving only a slight adjustment of the original machine. It is a far more complicated operation to sew the upper to the thick and heavy sole, and years passed by before the secret was discovered, and the McKay machine appeared. In the shoe sewed on the McKay machine, the thread ran through into the inside of the inner sole, leaving a rasping ridge on which the stocking of the wearer rubbed. The McKay shoe displaced only the coarser grades. The hand-sewed shoe remained the favorite of wealth and fashion, and was worn exclusively by those who cared for comfort and could afford the price. In sewing a shoe by hand, a thin and narrow strip of leather, called a welt, is first sewed to the insole and upper, and the heavy outsole is sewed to this welt, so that the stitches come outside and do not touch the foot, the insole being left entirely smooth. It is a delicate operation by hand, and many years elapsed before a machine was contrived by which it could be done. At last the problem was solved. The "Goodyear welting and stitching machines" appeared—so named for Charles Goodyear, who financed and perfected them, a son of the man who taught the world the use of rubber. These two machines are the nucleus of the Goodyear welt system, to which must be attributed the revolution of an industry. Although they are entirely distinct machines, they are inseparable, for neither can be used effectively without the other in making the modern Goodyear welt shoe.

Much of the style of a shoe depends upon the wooden last over which the upper is shaped before being attached to the sole. To find a substitute for the human hand in fitting the shoe to the last and pulling the leather over its delicate lines and curves seemed for a long time impossible.

This took place in the early seventies, when a machine was invented for doing this work. It created a great change in a department of shoemaking which, prior to this time, had been regarded as a confirmed hand process. This machine, as well as those which followed afterwards for a period of twenty years, was known as the best type of machine, by which the shoe upper was drawn over the last by either friction or pincers, and then tacked by use of a hand tool.

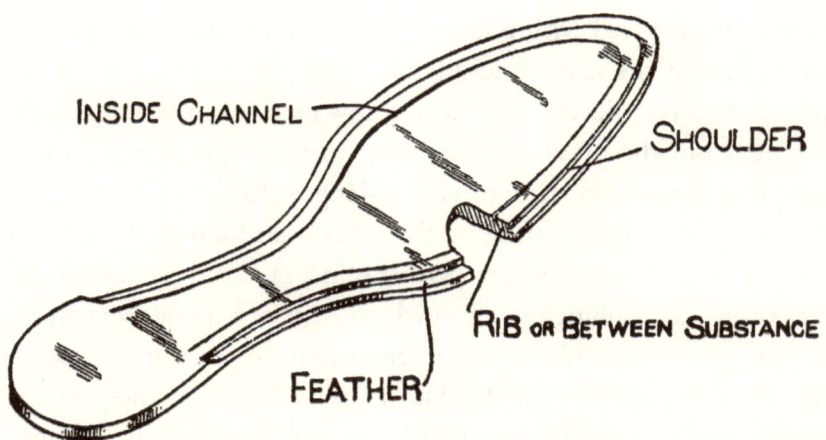

Insole for Hand Sewed Shoe.

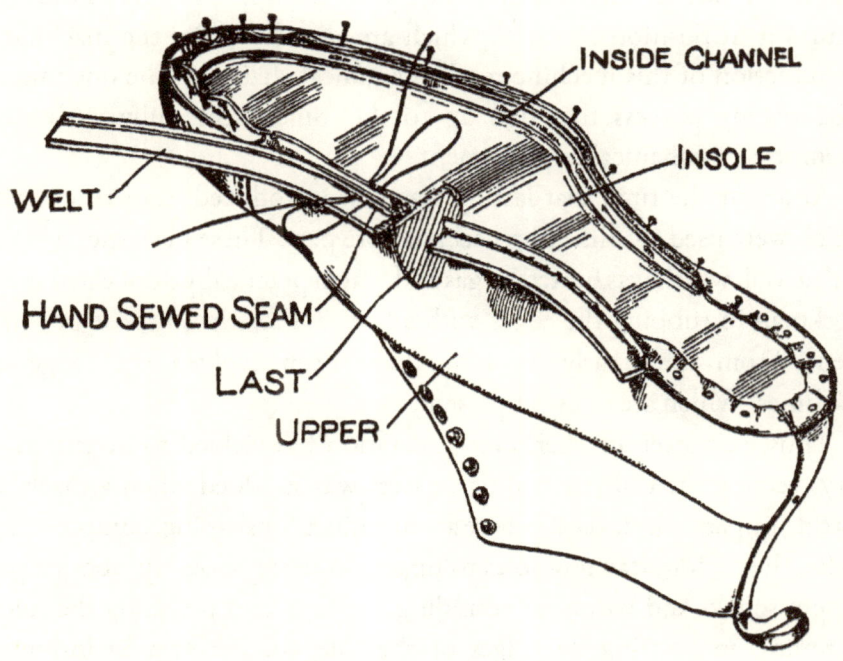

Hand Sewed Shoe.

At a comparatively recent period another machine which revolutionized all previous ideas in lasting was introduced. This machine is generally in use at the present time, and is known as the "consolidated hand method lasting machine." It was fitted with pincers, which automatically drew the leather round the last, at the same time driving a tack which held it in place. This machine has been so developed that it is now used for the lasting of shoes of every type, from the lowest and cheapest to the highest grade, and it is a machine that shows wonderful mechanical ingenuity.

The perfection of the lasting machine has been followed recently by the introduction of a machine which performs in a satisfactory way the difficult process known as "pulling over," which consists of accurately centering the shoe upper on the last and securing it temporarily in position for the work of lasting. The new machine, which is known as the "hand-method pulling over machine," is provided with pincers, which close automatically, gripping the shoe upper at sides and toe. It is fitted with adjustments by which the operator is enabled to quickly center the shoe upper on the last, and, on pressure of a foot lever, the machine automatically draws the upper closely to the last and secures it in position by tacks, which are also driven by the machine. The introduction of this machine marked a radical change in the one important shoe-making process that had up to this time successfully withstood all attempts at mechanical improvement.

At about the time that lasting was first introduced, came the machines which were used for finishing heel and fore part. These machines were fitted with a tool, which was heated by gas and which practically duplicated the hand workman in rubbing the edges with a hot tool for the purpose of finishing them. From these early machines have been evolved the "edge-setting machines" which are in use at present.

Thus, one after another, every operation has yielded to invention, until very recently the only remaining process was subdued when a machine for cutting uppers was devised. There are machines for shaping, compressing, and nailing heels; for attaching soles to uppers in heavy shoes by wooden pegs or copper screws and wires; for rounding, buffing, and polishing the soles; for trimming and setting the edges of the sole; for performing innumerable operations, some seemingly trivial, but all essential to perfection in comfort, durability or style; so that in shoe factories to-day a greater variety of intricate and expensive machines is used than in factories of any other kind.

At the present time the genius of the American inventor has provided for every detail of shoemaking, even the smallest processes being performed by mechanical devices of some kind. This has naturally made the shoemaker of to-day a specialist, who very seldom knows anything of shoe-making apart from the particular process in the performance of shoemaking of which he is an adept, and from which he earns a livelihood. The American shoe of to-day is the standard production of the world. It is in demand wherever shoes are worn.

In the year 1874 there had been perfected not only the machines which Colonel McKay and Mr. Goodyear had been instrumental in building, but other inventors had introduced similar machines for doing similar work. This brought about the most acute business competition, and finally resulted in many cases where one machine manufacturer alleged that the other machine infringed his rights of patent, and in many other cases the fiercest kind of litigation was established. This had a most disastrous effect upon shoe manufacturers, for in many cases the manufacturer was made to bear the brunt of the blows which contending shoe machinery manufacturers aimed at each other.

Machines in use in factories were stopped by means of injunctions; damage suits were entered, and litigation was very general. During the year 1899, there was ushered in one of the most important events that ever transpired in the history of shoemaking. The most important of the concerns which had been making war upon each other were purchased by one large company and brought under one harmonious management.

The United Shoe Machinery Company owes its origin to a call for a change in conditions menacing the industry of making shoes which could not be ignored. It was created by combining into one the three companies existing in 1899: the Goodyear Sewing Machine Company, the Consolidated & McKay Lasting Machine Company, and the McKay Shoe Machinery Company, each of which respectively made and leased machines adapted to a particular class of operations. The principal machines which each made did not interfere with the principal machines of any other. They were dependent links in an industrial chain. The Goodyear Sewing Machine Company chiefly made machines for sewing the sole to the upper in welt shoes and various auxiliary machines which helped to complete the shoe; The Consolidated McKay Lasting Machine Company made machines for lasting a shoe; The McKay Shoe Machinery Company made various machines for attaching soles and heels by metallic fastenings, and furnished material for that purpose. A single manufacturer, in order to make Goodyear welt shoes, would be compelled to patronize all the companies, going to each of them for that part of his equipment which it exclusively supplied. Each company had its agents in factories looking after its machines.

The gathering of these three companies into a single organization wrought an instant change. It resulted immediately in greater economy of administration; in relieving the manufacturer of the vexation of sometimes seeing his factory crippled while orders were piling up; in freeing him from the annoyance and expense of dealing with several different concerns in order to get his most important machines and keep them in repair.

The attention which had been paid to royalty machines and which had been such an important factor in building up the industry in America, was magnified by the management of the new company. Large forces of men and expert machinists, as well as expert shoemakers, were maintained in the different districts where shoes were made, and every effort exerted to promote the growth of the industry.

While the royalty system proved to be of great advantage to small shoe manufacturers, the largest manufacturers objected to paying royalty on machines and desired to purchase them outright. Being unable to do so, they placed experts at work to invent similar machines. This has resulted in the United Shoe Machinery Company claiming that these machines are infringements and causing considerable litigation.

If one reviews the history of the trade during the past ten years, there will be little question but that one will find it has been a period of the greatest advancement that the trade has ever known.

Within the time of those who read these words, the way to make a shoe has been completely changed. Methods which held their own for centuries have disappeared, to be replaced by processes which only recently would have been thought impossible, and which have brought within the reach of men of modest means a luxury once enjoyed exclusively by the well-to-do. The feet of the million are clad to-day as finely as the feet of yesterday's millionaire. Shoes marked by comfort, durability, and style have driven to historical museums the stiff and clumsy boots and brogans which not so many years ago were worn by those who could not pay to have shoes sewed by hand.

The American people spend more than three hundred million dollars every year in buying shoes, and average three pairs apiece, and yet few ever think about their shoes so long as they do not look clumsy, or wear out too quickly, or hurt the foot. Every one likes to buy good shoes as cheaply as he can, and every one likes to feel that shoe manufacturers are independent and successful, and that workmen get good wages, because these things help along prosperity; but that is all. Yet here is an industry in which the United States within a decade has come to lead the world, and there are many things about it which it would be worth while for every one to understand. It is worth while, for instance, to know that there is no important operation on a shoe which need be done by hand; that in the making of every good shoe no less than fifty-eight different machines, and sometimes twice that number, are brought into play; that nearly all these machines are of American invention; and that they have been so perfectly adjusted one to another that they work together almost with the precision of a watch; it is worth while to know something about the marvelous system under the encouragement of which this typical American industry has blossomed and borne fruit until it employs two hundred million dollars of capital and nearly two hundred thousand people, and turns out two hundred and fifty million pairs of shoes a year; and why it is that the average man you meet to-day has a better fitting, better wearing, and better looking shoe than the moneyed man of yesterday—at a fraction of the expense.

This remarkable growth is distinctly American. In the United States the tendency among the artisan class has been to abandon the slow hand process. This tendency has been as strong as the tendency in Europe to adhere to it. Moreover, there has developed among the laboring classes in the United States a mobility such as is unknown elsewhere in the world.

Another advantage which has contributed to the rapid development of the manufacture of shoes in the United States is the comparative freedom from inherited and overconservative ideas. This country has entered upon its industrial development unfettered by the old order of things, and with a tendency on the part of the people to seek the best and quickest way to accomplish every object.

In all of the European countries in which the manufacturing of shoes is an important industry, the transition from the household to the factory system was hampered by guilds, elaborate national and local restrictions, and by the national reluctance with which a people accustomed for generations to fixed methods of work, in which they have acquired a large degree of skill, abandon those methods for new ones. It was natural, also, that in spite of the superior advantages of machine methods, hand process of manufacture should still continue side by side with them, in the European countries, though machine work had long since usurped the whole field of the shoe industry in the United States.

Stitching Room of a German Shoe Factory.

As an American goes about among the European shoe factories he is greatly surprised at the state of affairs. He is struck by three things which are very conspicuous. They are: (1) Lack of use of machinery, lack of all sorts of devices in order to save hand labor, which is carried out so extensively in the United States. (2) Lack of the division of labor, one factory attempting to make four or five kinds of shoes. (3) Lack of methods employed for handling large quantities of materials.

One point that is overlooked in considering the shoe industries of the two countries is the great difference in organization. In most European factories, the manufacturer gets all the orders of different kinds, and then attempts to make one or two lines with one or two qualities in the same factory. In

Switzerland one may find shoes and slippers for men, women, and children made under the same roof.

In the United States the manufacturer makes a certain line of shoes in one factory, and no other kind. If he has more than one line, he has more than one factory, and each factory turns out a distinct shoe for a distinct purpose. The manufacturer has his salesmen to sell these shoes.

The advantages of the American system are: (1) The managers and workers of a factory turning out a certain line of goods become highly specialized in that line, and can produce better results than the workers in a factory attempting to make two or three lines of goods. (2) A large shoe factory is laid out as a rule to do a certain kind of work, and it seldom changes. This practice makes possible a greater production. On the other hand we have something to learn from the European organization. American manufacturers must meet the foreign trade. In order to do this, the manufacturer must cater to the habits, customs, and climatic conditions. The European manufacturer does this.

www.ingramcontent.com/pod-product-compliance
Lightning Source LLC
Chambersburg PA
CBHW020425010526
44118CB00010B/427